THE NORTH AMERICAN SEA DUCKS

The North American

Sea Ducks

Their Biology and Behavior

Paul A. Johnsgard

School of Biological Sciences
University of Nebraska–Lincoln

Zea Books, Lincoln, Nebraska 2016

*Dedicated to all the biologists whose often dangerous fieldwork has resulted
in our current knowledge of sea duck biology, and especially to the memory of those
who gave their lives in the effort.*

Abstract

The 21 species of sea ducks are one of the larger subgroups (Tribe Mergini) of the waterfowl family Anatidae, and the 16 species (one historically extinct) that are native to North America represent the largest number to be found on any continent, and also the largest number of endemic sea duck species native to any continent.

Although generally not important as game birds, the sea ducks include some economically important birds such as the eiders, the basis for the Arctic eiderdown industry and a historically important food source for some Native American cultures. They also include what is probably the most northerly breeding species of all waterfowl and an icon of Arctic bird life, the long-tailed duck. The sea ducks also include species having some of the most complex and diverse pair-forming postural and acoustic displays of all waterfowl (goldeneyes and bufflehead), and some of the deepest diving species of all waterfowl (scoters and long-tailed duck). Sea ducks are highly prone to population disasters caused by oil spills and other water contaminants and, like other seabirds, are among the first bird groups that are being affected by current global warming trends in polar regions.

This book is an effort to summarize succinctly our current knowledge of sea duck biology and to provide a convenient survey of the vast technical literature on the group, with over 900 literature references. It also includes 90,000 words of text (more than 40 percent of which is new), 15 updated range maps, 31 photographs, over 30 ink drawings, and nearly 150 sketches.

Lastly, the North American sea ducks include the now extinct Labrador duck, the only northern hemisphere waterfowl species to have gone extinct in modern times. I have gratefully reprinted a Labrador duck watercolor by Sir Peter Scott. Considering recent population crashes in other sea ducks, such as the Steller's eider and spectacled eider, it should also offer a sobering reminder of the fragility of our natural world and its inhabitants, including us.

ISBN 978-1-60962-106-3

http://doi.org/10.13014/K22Z13FS

Composed in Adobe Garamond and Imprint MT Shadow types.

Zea Books are published by the University of Nebraska–Lincoln Libraries
Electronic (pdf) edition available online at http://digitalcommons.unl.edu/zeabook/
Print edition available from http://www.lulu.com/spotlight/unllib

UNL does not discriminate based upon any protected status.
Please go to unl.edu/nondiscrimination

Nebraska
UNIVERSITY OF
Lincoln

Contents

Maps

The North American breeding and wintering ranges of the

Figures

Photographs

Preface

The sea ducks have been a favorite bird group of mine ever since childhood, although I grew up more than a thousand miles from either coast. It wasn't until my graduate student years that I was first able to observe sea ducks in their natural marine environment, and three decades would pass before I had seen all 15 surviving North American sea duck species in life, and all under natural conditions.

The very geographic remoteness and my little understanding of sea ducks have indeed been major sources of the appeal that I have long felt for sea ducks. Watching spectacled eiders and emperor geese flying over and courting on just-thawed tundra-rimmed wetlands beside the Bering coast of western Alaska, encountering families of Barrow's goldeneyes and harlequin ducks foraging in remote mountain ponds and navigating rapids in clear Rocky Mountain streams, and hearing the wild, evocative calls of long-tailed ducks echoing over the tundras of Arctic Canada are among my greatest joys and most cherished memories of my life.

I first wrote about all these enchanting birds in my *Waterfowl of North America* in 1975, and then again in 1978 in my *Ducks, Geese, and Swans of the World*. Both of these books are now about four decades old, and two new generations of waterfowl biologists have appeared and added their contributions to the already large literature on the group. To add to my 2016 updating of the swans, geese, and grouse of North America, I recently decided to undertake a revision of my earlier writings on the North American sea ducks, primarily consisting of additions, updates, and modifications of my *Waterfowl of North America*.

In this effort I have mainly tried to fill in most of the gaps in basic knowledge of sea duck biology that still existed in the 1970s. I have also updated all my range maps to varying degrees, but—with the exception of the Barrow's goldeneye map—I was unable to incorporate the boundaries of Nunavut, the territory established in 1999 that encompasses much of Arctic Canada. I have also updated species taxonomies and descriptions, added new drawings and photos, and incorporated about seven hundred new literature citations into those that were in my earlier waterfowl books. However, I have not added more information on topics that I felt I had previously summarized adequately for these species.

All the photographs, drawings, and maps are my own. My somewhat primitive sketches of sea duck displays were based on studies of 16mm cine frames and have been reprinted from my 1965 *Handbook of Waterfowl Behavior*. That book provides written descriptions of all the displays I have illustrated and can be freely accessed at http://digitalcommons.unl.edu/bioscihandwaterfowl/7/.

Two major reference books on waterfowl have been published since 1975, including a two-volume world survey of waterfowl, *Ducks, Geese, and Swans*, which was the marvelous swan-song effort by a long-time friend, Dr. Janet Kear (1933–2004), and which was published in 2005, shortly after her untimely death. Additionally, Professor Guy Baldassarre (1953–2012) provided a splendid major revision of F. H. Fortnight's classic 1943 *Ducks, Geese, and Swans of North America*, which was likewise, sadly, published posthumously in 2014.

Important individual species monographs by various authors of all the North American sea ducks have also appeared since the 1990s, through the joint multiyear *The Birds of North America* project of the American Ornithologists' Union and the Philadelphia Academy of Natural Sciences, which has recently become available online through the cooperation of Cornell University's Laboratory of Ornithology (the Birds of North America Online, https://birdsna.org/Species-Account/bna/home).

These valuable sources collectively provide far more comprehensive information than I could incorporate into the present work. For example, I have excluded detailed molt and plumage descriptions as well as information on fossils, diseases, parasites, effects of pollution, and increasingly serious environmental contaminants, such as lead, mercury, and chlorinated hydrocarbons. However, to partly compensate I have tried to assemble a much larger and more comprehensive bibliography of the North American sea ducks than was included in any of my earlier waterfowl books.

Several important research findings involving sea ducks have emerged since my 1975 book was published. The previously unknown wintering grounds of the spectacled eider have at last been discovered far out in the Bering Sea, and the many advances in radar techniques, satellite tracking, and radiotelemetry have greatly increased our knowledge of sea duck movements, migrations, and mortality. Finally, much more has been learned about the reproductive biology of such little-known species as the Steller's eider, spectacled eider, and surf scoter.

Sadly, many sea ducks, especially the coast-loving eiders and scoters, have undergone serious population declines since 1975. Increasing human population influences, such as widespread overexploitation of fish and shellfish, oil spills, water pollution, ingestion of lead, mercury, selenium, and other chemical poisons, and breeding habitat destruction have certainly been contributing factors in these population declines. Global warming is also progressively affecting ocean temperatures, sea levels, coastal breeding and wintering habitats, and continental weather patterns, with increasingly destructive and potentially fatal ramifications ahead for both humans and wildlife.

I have already acknowledged many of the people who helped me with the earlier editions of my waterfowl books, but I wish to again acknowledge the memory of Sir Peter Scott (1909–1989) who, together with his staff and the facilities of England's Wildfowl Trust (now the Wildfowl and Wetlands Trust), truly enabled me to launch my career as a waterfowl biologist more than 50 years ago. Like Janet Kear—and many other great biologists who personally touched and enriched my life—Peter Scott has now sadly left us, but his unsurpassed waterfowl paintings still survive, and it seems appropriate that I should reprint the evocative watercolor of the likewise now-vanished Labrador duck that he kindly executed for my *Waterfowl of North America*.

Additionally, as with my many previous monographs that have already been placed into the UNL DigitalCommons library, I owe a huge debt of gratitude to Paul Royster, Coordinator of Scholarly Communications for the University of Nebraska–Lincoln Libraries, and to his editorial staff for seeing this book through to completion. Thanks also to the university's librarians for their cheerful help in locating and providing reference materials for me, and to the University of Nebraska for continuing, for still unfathomable reasons, to support my research activities.

Paul A. Johnsgard
Foundation Regents Professor Emeritus
Biological Sciences
University of Nebraska–Lincoln

Hooded merganser, male crest-raising

I. The Sea Ducks of the Tribe Mergini

The sea duck tribe Mergini, which includes the eiders, mergansers, scoters, goldeneyes, and a few other unique taxa, consists of 21 species of waterfowl, all having strong swimming and superb diving abilities. Although called "sea" ducks, many species are likely to be found in both fresh-water and marine habitats, and a few, such as the hooded merganser, might spend their entire lives in freshwater environments. All but two of the sea duck species have entirely Northern Hemisphere distributions and all of these undergo seasonal migrations of greatly varying distances.

Of the 21 currently recognized species of sea ducks, North America is represented with 15 surviving breeding species. Most of the North American species also occur extensively in the Old World, with the exception of the surf scoter, bufflehead, Barrow's goldeneye, and hooded merganser. Additionally, the American Ornithologists' Union (AOU) has recently recognized the North American black scoter (*Melanitta americana*) as a distinct species from the very similar Eurasian common scoter (*Melanitta nigra*), adding a fifth species to the list of endemic North American sea ducks.

One other North Americans endemic sea duck, the Labrador duck, became extinct in the late 1800s, before its biology and natural history had become known, and thus is only briefly described here. The smew (*Mergellus albellus*) is a small Eurasian merganser that is extremely rare in North America. Most of the authenticated sightings of smews have been winter records from Alaska and the Pacific and Atlantic coasts. The smew was briefly described in my 1975 book but has been excluded from this book because of its rarity and the absence of any known North American breeding. As of the early 2000s there were at least five records for British Columbia, at least two for Washington and California, and one for Oregon.

All of the 15 surviving North American sea ducks are migratory to varying degrees, and many undertake long postbreeding "molt migrations" in late summer to isolated and secure habitats rich in food, where they undergo their annual wing molts and vulnerable flightless periods prior to wintering elsewhere. Most sea ducks winter offshore in marine environments, but some might also overwinter in coastal estuaries, and a few are prone to spend the winter in freshwater habitats.

All the sea ducks require at least two years to attain sexual maturity and acquire their definitive breeding plumages, and in some cases three or even four years might pass before initial reproduction occurs. Most sea ducks are relatively long-lived, sometimes surviving 10–15 years in the wild. Based on current evidence, monogamous pair-bonds are present in all species and are typically renewed annually, although there is evidence for occasional multiyear monogamous pair bonding in at least the genera *Bucephala*, *Clangula*, *Histrionicus*, and *Mergus*.

Since adult males always outnumber adult females, spirited completion for mates and prolonged, often complex and species-specific courtship display rituals are typical prior to pair-bonding. In correlation their annual courtship rituals, they exhibit marked sexual differences in adult plumages (sexual dichromatism) and sizes (sexual dimorphism), with the larger males often exhibiting conspicuous, sometimes colorful and partially iridescent breeding plumages, followed by a reversion to subdued female-like plumages during the nonbreeding season.

Except for two geographically isolated Southern Hemisphere species, the now-extinct Auckland Islands merganser (*Mergus australis*) and the critically endangered Brazilian merganser (*Mergus octosetaceus*), males of all of the world's sea ducks are mostly elaborately patterned in black and white, which probably serves very well to localize and identify the ages and sexes of these birds at considerable distances. However, iridescent plumage coloration of most North American sea ducks is generally limited to the heads of adult males.

Sea ducks usually winter in coastal waters, and most breed in tundra situations or in northern temperate forests. All 19 of the world's extant sea duck species depend predominantly on animal sources of food, and some feed exclusively on these materials. Their foods include crustaceans, mollusks, echinoderms, annelids, other invertebrates, and some aquatic vertebrates, especially fish. The sea ducks are thus not generally regarded as highly for table fare as are the surface-feeding ducks and some of the more plant-eating diving ducks.

In common with typical freshwater diving ducks such as the pochards (*Netta* and *Aythya*, tribe Aythyini), the webbed feet of sea ducks are unusually large, with long outer toes and well-lobed hind toes, and their legs are located close to the tail. The sea ducks have thus sacrificed the ability to walk easily in favor of more efficient diving adaptations, such as an oarlike nearly horizontal leg-stroking ability. Some species, especially the fish-catching mergansers, dive almost effortlessly and swim rapidly without opening their wings. Others, such as scoters, use their wings to assist in diving and might keep them partly extended for underwater steering while foraging.

In common with pochards, the generally heavier bodies of sea ducks relative to their wing surface area prevent them from taking flight without running some distance over the water prior to reaching minimum flight speed. Livizey (1995) attributed their large body mass with associated increased capabilities to dive deeply for benthic prey, as in the larger eiders, although the record-holder for deep diving (reportedly to at least 150 feet) might be held by the benthic-foraging long-tailed duck.

In flight, most sea ducks often make up in speed for their limited aerial maneuverability, although some of the larger eiders and scoters are rather ponderous in flight. However, mergansers are notable for their streamlined body forms and high flight speeds (reportedly in excess of 100 miles per hour by a red-breasted merganser that was being chased by an airplane).

Most sea ducks exhibit a substantial amount of white on the wings while in flight, and, like many surface-feeding ducks, two species (the Steller's eider and harlequin duck) exhibit iridescent speculum patterns on their secondary flight feathers. The bills of breeding males are often brightly colored and some are distinctively shaped—these might be important in species and sexual recognition among scoters and eiders—and the feet and legs are also often notably colorful among adult male sea ducks.

Many male sea ducks utter distinctive vocal calls or whistles during courtship (oldsquaws, eiders), or produce whistling sounds mechanically through outer wing feather vibrations during flight (scoters, goldeneyes), but female sea duck voices are notably subdued and infrequent. Male displays range from a few rather simple postures, as in scoters, to diverse and highly stereotyped ("ritualized") movements, as in male eiders and goldeneyes. Female inciting behavior, her encouragement of a preferred male to threaten or attack other rivals, and a variably prolonged and motionless prone precopulatory posture that is assumed prior to male mounting, while the male performs species-typical preliminary displays, are typical traits of sea ducks.

Nesting by sea ducks might be done on fairly open shoreline or in grassy tundra, as is typical of eiders and the long-tailed duck, or in wooden or other natural cavities, under or amid thick vegetation, or in unusually well-concealed locations, as are found among goldeneyes and mergansers. The Arctic-breeding and tundra-nesting sea ducks such as eiders typically build open-cup nests in low tundra vegetation. Females somewhat resemble female dabbling ducks in their cryptic plumage patterns, and their downy young are also rather obscurely patterned. These ground-nesting sea ducks usually place their nests fairly near the water's edge, such as along shorelines or on small islands, although the sometimes inland-nesting scoters are an exception to this generalization.

In contrast, the forest-nesting sea ducks often use hollow trees, crevices, exposed tree roots, or other natural cavities for their nest sites. Females of these cavity nesters are more uniformly colored than are surface nesters, but their ducklings are often contrastingly patterned with white and dark markings. Some of the tree-nesting sea ducks have relatively long, broad tails that assist in tree landings and can perch fairly well. Cavity-nesting sea ducks also show a higher degree of year-to-year nest-site fidelity than do ground nesters, their clutch sizes tend to be larger, and both their incubation and fledging periods are often somewhat longer.

The relatively enclosed condition of cavity nests makes an effective incubation of fairly large clutches more feasible, and their longer incubation periods are often attributed to their reduced dangers of predation and random egg losses. One of the costs of cavity nesting, however, is increased competition for these relatively rare sites, and the consequent probability of alien eggs being insinuated into the owner's clutch by another bird ("egg-dumping" or "brood parasitism").

Brood parasitism, both within and between species, is a common practice among cavity-nesting ducks, and sometimes has significant effects on hatching success for the host. Besides often having somewhat larger clutch sizes than surface nesters, cavity-nesting sea ducks also tend to lay relatively round eggs, which take up less nest space relative to their volume than do oval eggs and because of the confined space in the nest are in little or no danger of falling out.

Whether of cavity-nesting or surface-nesting birds, clutches of sea duck eggs tend to hatch more or less simultaneously, allowing the females to leave the nest with her entire brood, often within 24 hours of hatching. Brood amalgamation, in which the young from several different families merge to form crèches that are tended by one or more hens, is common in some sea ducks, notably eiders and scoters.

In contrast, goldeneye females are notably possessive of their broods, and in their efforts to evict intruding ducklings might mistakenly harm or even kill one or more of their own offspring. Among a few sea ducks, such as some mergansers and goldeneyes, the downy young sometimes climb up, and might even ride, on the back of their mother while she is resting or swimming. This behavior is seemingly not encouraged by the hen, but the location provides a warm and dry if temporary environment for one or more offspring.

The sea ducks have been recognized as a natural group of related species ever since 1945, when they were defined as a unique waterfowl subgroup (tribe Mergini) by Jean Delacour and Ernst Mayr in their classic revision of the waterfowl family Anatidae. Twenty years after my 1975 publication of *Waterfowl of North America*, Bradley Livizey (1995) undertook a revision of the sea ducks, using cladistics to estimate the phyletic relationships within this group. Like me, he concluded that the eiders are a sister group to the other Mergini, as was first envisioned by Delacour and Mayr.

Livezey suggested a linear taxonomic sequence of sea duck genera as follows: *Polysticta, Somateria, Histrionicus, Camptorhynchus, Melanitta, Clangula, Bucephala, Mergellus, Mergus*. This sequence has been adopted by the American Ornithologists' Union (AOU) and is very close to my proposed generic sequence of the family Anatidae (Johnsgard, 1979, 1997). It differs from mine mainly in the position of *Clangula* and convinced me to leave my proposed sequence unchanged for this book.

It is sad that two of the sea ducks—the Labrador duck and Auckland Islands merganser—are now extinct, and several others—such as the spectacled eider and Steller's eider—currently have dangerously low populations. The loss of the two extinct species occurred more than a century ago, with no obvious attempts having been made to prevent or even to try to understand the reasons for their loss.

The Labrador duck was thus added to the list of the many conservation failures of North Americans over the last two centuries, which also included the extinctions of the Eskimo curlew, passenger pigeon, Carolina parakeet, and ivory-billed woodpecker, and the current critically endangered status of several others. The beautiful cover watercolor illustrating a flock of Labrador ducks that was painted by the late Sir Peter Scott can only increase our collective sense of sadness for the absence of such a spectacular bird from our lives, and serve as a harbinger of the loss of biological diversity that the world has since increasingly suffered.

II. Species Accounts

Red-breasted merganser, adult male

Common Eider
Somateria mollissima (Linnaeus) 1758

Other vernacular names. American eider, northern eider, Pacific eider

Range. Breeds in a circumpolar distribution across Greenland, Iceland, the British Isles, Scandinavia, Novaya Zemlya, Russia, coastal Siberia, and Kamchatka; and in North America from the Aleutian Islands and the Alaska Peninsula to western and northern coastal Alaska, the Arctic coast of Yukon Territory, Northwest Territories, Nunavut Territory, and offshore Arctic islands; also Hudson Bay, Labrador, Newfoundland, the Gulf of St. Lawrence, New Brunswick, Nova Scotia, Maine, and Massachusetts. In North America, winters in coastal areas of the Pacific south to Washington and along the Atlantic coast south to the middle Atlantic states, with casual occurrences inland.

Subspecies. *S. m. dresseri* Sharpe. American Common Eider. Breeds in southern Labrador, Newfoundland, the Gulf of St. Lawrence, Nova Scotia, Maine, and Massachusetts.

S. m. borealis (Brehm). Northern Common Eider. In North America breeds from Greenland and northeastern Canada west to northern Hudson Bay, where it probably intergrades with *sedentaria*.

S. m. sedentaria Snyder. Hudson Bay Common Eider. Resident in Hudson Bay from Cape Fullerton to the east coast south of Southampton and Coats Islands, and south into James Bay.

S. m. v-nigrum Bonaparte. Pacific Common Eider. In North America, breeds from northern Alaska east along Canada's coastal Northwest Territories, Banks Island, Victoria Island, and south to the Aleutian Islands, Kodiak Island, and along the southern coast of the Alaska Peninsula east to Cook Inlet and Glacier Bay. Also breeds in northern and Siberian Russia, and on St. Lawrence Island.

Measurements. *Folded wing: S. m. borealis:* ave. of 147 males 52.4 mm, of 169 females 47.7 mm; *S. m. dresseri:* ave. of 16 males 287 mm, of 21 females, 277 mm; *S. m. sedentaria:* ave. of 4 males 311 mm, of 18 females 303 mm; *S. m. v-nigrum* ave. of 22 males 303 mm, of 20 females 292 mm (Goudie, Robertson, and Reed, 2000).

Culmen (along midline): *S. m. borealis:* ave. of 12 males 57.8 mm, of 16 females 52.9 mm; *S. m. dresseri:* ave. of 16 males 50.4 mm, of 21 females 50.0 mm; *S. m. sedentaria:* ave. of 9 males 58.1 mm, of 161 females 52.2 mm; *S. m. v-nigrum:* ave. of 25 males 49.4 mm, of 20 females 52.7 mm (Goudie, Robertson, and Reed, 2000).

Weights (mass). Goudie, Robertson, and Reed (2000): *S. m. borealis:* ave. of 28 females 1,648 g. *S. m. sedentaria:* ave. of 4 males 2,276 g; of 18 females 2,147 g.

Nelson and Martin (1953): *S. m. dresseri:* 8 males ave. 4.4 lb. (1,995 g), max. 4.6 lb. (2,084 g); 8 females ave. 3.4 lb. (1,542 g), max. 3.8 lb. (1,721 g). *S. m. borealis:* 1 male 3.4 lb. (1,542 g); 9 females ave. 3.4 lb. (1,542 g). *S. m. v-nigrum:* 8 males ave. 5.7 lb. (2,585 g), max. 6.2 lb. (2,809 g); 4 females ave. 5.4 lb. (2,449 g), max. 6.4 lb. (2,898 g).

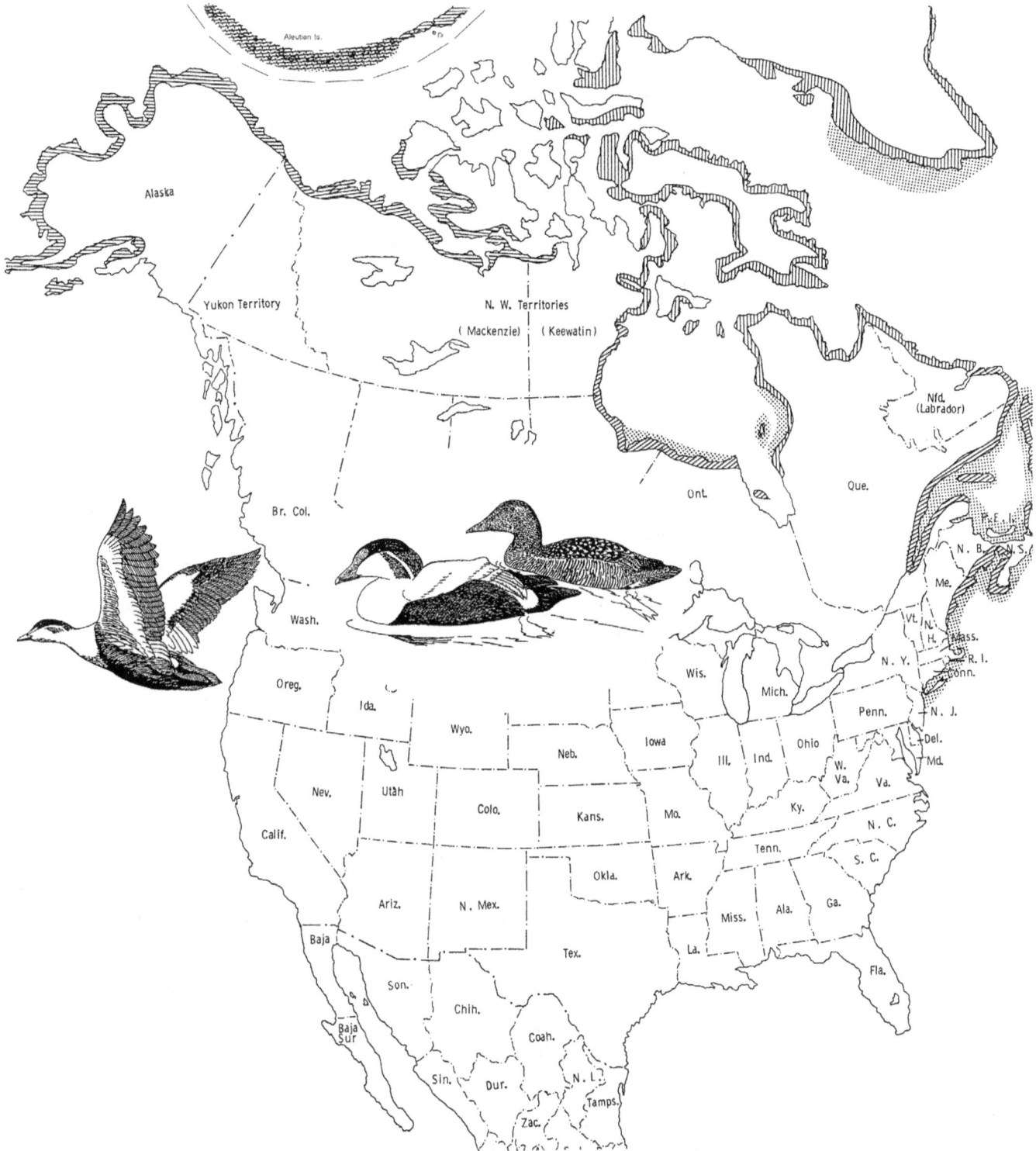

The North American breeding (hatched) and wintering (shaded) ranges of the common eider. Horizontal hatching indicates breeding ranges of Pacific common eider, vertical hatching the northern and Hudson Bay common eiders, and diagonal hatching the American common eider.

Identification

In the hand. In the hand, specimens can be immediately recognized as eiders by the somewhat sickle-shaped tertials and the irregular basal feathering of the bill, plus the rather large body that usually weighs in excess of 3 pounds. The common eider differs from all other eiders in having a lateral extension of feathering on the side of the bill that tapers to a point below the rear tip of the nostrils along with an unfeathered extension of the bill that extends nearly to the eyes; these are present in both sexes and all ages. The bill color and the width of the unfeathered extension toward the eyes varies with different subspecies. For adult males, the presence of white or mostly white tertials and a folded wing length in excess of 270 mm separates common eiders from spectacled eiders and king eiders.

In the field. On the water, common eiders can be recognized at great distances by the male's white mantle color, which extends downward on the breast to the anterior base of the wings. King eiders have a black mantle, and spectacled eiders have blackish color extending partway up the breast toward the front of the neck. Female common eiders are less rusty-toned and paler than female king eiders and are vertically barred with dark brown rather than having crescent-shaped brown markings. In flight, common eiders travel in a straight course with strong wing strokes; the males exhibit a continuous white mantle that connects their white upper wing-coverts, and they have a black crown-stripe that is lacking in the other eiders. During courtship, male common eiders utter rather loud cooing sounds, similar to those of mourning doves but lacking the tremulous quality of corresponding king eider calls. Female calls are fairly loud and hoarse, often sounding like *gog-gag-gog*, and lack the wooden tone of king eider calls.

Age and Sex Criteria

Sex determination. By the first spring of life, male eiders will have acquired at least some white feathers on their breasts, which together with the white upper wing-coverts should be evident in any plumage. Even in juvenile plumage, the male has a lighter chest than the female.

Age determination. Age criteria for females have not been worked out, but juveniles of either sex should be recognizable by the notched tail feather criterion. Older females should be examined internally as to the state of their reproductive organs, while males can probably be aged according to the distinctions mentioned for first-year and second-year king eider males. Reproductive maturity probably occurs during the second spring of life for females, although males of the same age still retain gray coloration on their upper wing-coverts.

Distribution and Habitat

Breeding distribution and habitat. In North America the common eider has the most extensive breeding range of any of the four eider species, from the Aleutian Islands on the west to Newfoundland on the east,

Common eider, male in breeding plumage

and from about 43° to 80° N. latitude (Maine to Ellesmere Island). In Alaska, Pacific common eiders nest on the Aleutian Islands, Kodiak Islands, the adjacent Alaska Peninsula eastward to Cook Inlet, northward in coastal tundra along the Bering coast, and on Nunivak and St. Lawrence Islands. They also breed from Tigara and Wainwright along the Arctic coast westward to Demarcation Point (Gabrielson and Lincoln, 1959).

In Canada the species breeds along the Yukon and Mackenzie coastline eastward to at least Bathurst Inlet, from the Melville Peninsula southward along the coastline of Hudson Bay to James Bay, along the eastern shores of Hudson Bay, and on the Atlantic coastline of eastern Canada to the mouth of the St. Lawrence River. It also breeds in Newfoundland and the other Maritime Provinces. In the Franklin District its breeding range includes the coastlines of Baffin and Southampton Islands, the smaller Hudson Bay islands, and at least parts of Banks, Victoria, Somerset, Cornwallis, Devon, and Ellesmere Islands (Godfrey, 1986). Victoria Island and Bathurst Inlet evidently represent the eastern limits of the Pacific race, and there is seemingly a hiatus between the breeding range of this form and the more easterly races.

South of Canada, the only states having native breeding eiders are Maine, New Hampshire, and Massachusetts. Maine has long had nesting eiders (Gross, 1944) and has also had remarkable population increases in recent years. Thus, the Muscongus Bay population in Maine rose from an estimated 800 birds in 1949 to more than 7,000 in 1965. By 1967 the Maine population had reached about 20,000 pairs (Mendall,

1968). Since then the population has seemingly stabilized at about 25,000 pairs. During the 1970s eiders from the Maine population were introduced on Penekese Island in Massachusetts, which have since spread to the Elizabeth Islands, Bird Island, and possibly others (Veit and Petersen, 1993).

The preferred breeding habitat of common eiders consists of low-lying rocky marine shores in the presence of numerous islands; there is also a rare utilization of sandy islands and coastal freshwater lakes or rivers (Snyder, 1957). Hildén (1964) found the highest nesting abundance on boulder-covered islands, with very little use of those that were gravel or rock covered. He also reported that grassy islands have highest usage, followed by those covered with herbaceous and, lastly, wooded vegetation. Most eiders selected central parts of islets for nesting rather than the shoreline area, perhaps as a reflection of their adaptation to tidal changes and also the scarcity of fine soil between the rocks to serve as nest substrates. Open terrain with extensive water areas, sparsely wooded islands with barren shores, as well as a proximity to marine foods are basic aspects of its habitat requirements.

Wintering distribution and habitat. The Pacific common eider winters throughout the Aleutian Islands and in the Bering Sea, where they are sometimes abundant around the Pribilof Islands and occasionally at St. Lawrence Island (Gabrielson and Lincoln, 1959). Fay (1960) stated that upward of 50,000 eiders (both Pacific common and king eiders) winter about St. Lawrence Island, although the king eider is much commoner and the Pacific eider is most prevalent during spring and summer months. There is probably also a western movement of birds from northwestern Canada around northern Alaska to wintering areas in the Bering Sea (Godfrey, 1986).

The remaining North American races of the common eider all winter in eastern Canada and along the Atlantic coast of the United States. In Canada wintering birds occur from southern Baffin Island and the islands north of Hudson Bay southward into Hudson and James Bays and eastward to the Maritime Provinces (Godfrey, 1986). Probably most Canadian eiders winter in the waters off southern Greenland and the Labrador coast (Snyder, 1957). Regular wintering south of Canada occurs along the coast of Maine, although birds sometimes move as far south as New York and New Jersey, with casual occurrences as far south as North Carolina.

In winter common eiders are almost strictly marine birds, usually remaining well offshore and generally out of sight from land. No doubt the availability of their winter foods (mollusks and some crustaceans), as determined by their abundance and water depth, are primary aspects of wintering habitat.

General Biology

Age at maturity. Data by Baillie and Milne (1982) indicate that some (26 percent) female common eiders first breed in Scotland when two years old, more (42 percent) when three, and still more at four years. Males probably don't breed until they are at least three years of age (Goudie, Robertson, and Reed, 2000).

Pair-bond pattern. Pair-bonds are normally renewed annually during the winter or spring months, although Spurr and Milne (1976) reported that remating with the same mate might sometimes occur over subsequent years. Males desert their mates early in the incubation period and often then directly migrate out to sea (Coach, 1965). Since many, but not all, birds are paired on their arrival, it would seem that pair formation might occur during migration or on the breeding grounds, according to Cooch. Kenyon (1961) noted that in the Pacific common eider pairing occurred in early or mid-May, about a month prior to the start of egg-laying. Thus, the pair bond of at least some eiders might last no more than a month or two.

Nest location. Coach (1965) reported that at Cape Dorset the northern common eiders favored nesting areas sheltered by rocks over flat, open, and grassy areas by a ratio of about 9:1. They also often placed their nests under rock overhangs, and they tended to select ridges that were well drained and normally were snow-free early in the season. About 40 percent of the nests found were within 100 feet of water, but at least 10 percent were more than 900 feet from water, so that immediate proximity to water is not necessary. Hildén (1964) similarly found that the birds were about ten times as abundant on boulder-covered islets than on either gravelly or rocky islets, and they favored those dominated by grassy rather than herbaceous or wooded vegetation. In his area, the eiders were not socially attracted to gulls or terns, but in a Spitsbergen study it was reported that nesting in association with Arctic terns (*Sterna paradisaea*) increased nesting success (Ahlen and Andersson, 1970).

Clutch size. Although average clutch sizes of 4.5 to 5.5 eggs appear to be typical of common eiders in both Scotland (Marshall, 1967) and Finland (Hildén, 1964), those of populations in eastern North America average between 3.25 and 4.04 eggs (Paynter, 1951). The modal clutch size of the Pacific common eider also appears to be 4 eggs (Kenyon, 1961), with a maximum of 6. Cooch (1965) similarly reported an average clutch size of 4.06 eggs for 188 first nestings of the northern common eider, as compared with 2.33 eggs in 12 renesting attempts. Eggs are normally laid at the rate of 1 per day, according to Cooch. Guignion (1969) reported an average of 4.3 eggs for completed clutches of the American common eider, and Freeman (1970) found that in the Hudson Bay population 536 nests averaged 4.5 eggs.

Incubation period. Cooch (1965) estimated a 28- to 30-day incubation period for northern common eiders in the wild. Guignion (1969) established that the American common eider had a 25- to 26-day incubation period under natural conditions, with nests in less disturbed areas hatching in a day less than those subject to some disturbance. Under artificial conditions incubation requires 24 to 25 days (Rolnik, 1943).

Fledging period. Cooch (1965) estimated a 60-day fledging period for northern common eiders. Goudie, Robertson, and Reed (2000) reported a 60- to 65-day period. Other fledging estimates by workers in Europe and Asia have ranged from 60 to 75 days (Bauer and Glutz, 1969).

American common eider, three males courting a female

Nest and egg losses. Cooch (1965) reported that at Cape Dorset, predation and other losses accounted for 25 percent of all eggs laid during one season and 15 percent the following one. However, there was some renesting, which tended to offset these losses. Three avian predators—ravens, herring gulls, and parasitic jaegers—were present on the area, but egg losses caused by jaegers were believed negligible. Hildén (1964) determined a similar nesting success of 78 percent in Finland, which was lower than that of any other waterfowl species in his study area. He attributed the high rate of nest failure, which was primarily caused by crows, ravens, and human interference, to the eiders' exposed nests, their failure to rapidly return after being flushed from the nest, the tendency to desert nests, and early initiation of nesting.

Choate (1967) reported that nest predation caused losses of 58 percent of 448 nests in Maine, over half of these losses occurring on incomplete nests. He found that larger clutches rather than smaller had a greater chance of success, which was considered to be possibly related to the age of the nesting female or her relative attachment to the clutch. Nests under cow parsnip (*Heracleum*), which provided cover for 426 of 963 nests, or under shrubs had higher hatching success than those placed in grasses or nightshade (*Solanum*). Gulls, including great black-backed and herring, that had caused partial predation on a clutch were often found to return and complete its destruction. An overall hatching success of 39 percent was found during

American common eider, males in breeding plumage

each of two years of study. On Spitsbergen (Svalbard), Arctic foxes and glaucous gulls were responsible for very high rates of nest and egg destruction (Ahlen and Andersson, 1970).

Summarizing data from eight studies, Goudie, Robertson, and Reed (2000) reported hatching success ranging from 25 to 80 percent (five studies), nesting success from 9.8 to 93 percent (five studies), and fledging success from 0.34 to 8.2 percent (two studies).

Juvenile mortality. Little information is available on prefledging mortality of young, largely because of brood merger, which is so prevalent in this species. Cooch (1965) believed that adverse weather, disease, and predation by various avian and mammalian predators might all play roles in determining juvenile mortality rates. Gulls (herring and glaucous) and ravens were observed chasing or attacking ducklings, especially during the first week after hatching.

Hildén (1964) estimated that over a three-year period and from a total of 1,026 eggs in Finland nests, 773 ducklings hatched and only 208 young survived to the end of the brood season. Single broods seemingly suffered fewer losses than did combined broods, although such brood merger has generally been considered an adaptation against gull predation. The great black-backed gull was evidently the worst duckling predator in Finland, but other causes of duckling mortality appeared to be disease, parasites, and, to a limited extent, weather. Hildén found that the number of young surviving until late in the brood season ranged in different years from 0.1 to 2.5 young per pair, or from 1.3 to 3.0 per female that he observed escorting ducklings. Estimates of postfledging mortality of juveniles would be very difficult to obtain in a marine habitat and are still unavailable.

Adult mortality. In North America, the annual survival of adult female American common eiders banded over ten years in the St. Lawrence estuary was 83 percent (Reed, 1975b). while for others banded over seven years in Maine and the Maritime Provinces the annual survival rate was 87.3 percent (Krementz et al., 1996). There are several examples of both sexes surviving in excess of 20 years in the wild.

General Ecology

Food and foraging. The importance of bivalve mollusks, especially the common blue mussel (*Mytilus edulis*), has long been recognized as a fundamental feature of the diet of common eiders. Dementiev and Gladkov (1967) reported that this species of *Mytilus* occurred in a sample of stomachs from Russian (eastern Murman) birds at a frequency of 70.3 percent, compared with 40.5 percent for *Balanus* barnacles and 24.3 percent for *Littorina* mollusks. In these and other samples various crustaceans (amphipods), echinoderms (sea stars and sea urchins), and fish (sticklebacks) occurred. Summer samples of juveniles and females showed amphipods, univalve mollusks, and ripening crowberries (*Empetrum*) present. Apparently periwinkles (*Littorina*) are a prime source of food for young ducklings.

A large sample of eiders taken between October and February in Danish waters supports the general view that mollusks (such as the bivalve *Mytilus* and the univalve periwinkle *Littorina*), crustaceans (especially the crabs *Carcinus* and *Balanus*), and sea stars (*Asterias*) are predominant parts of the winter diet of common eiders (Madsen, 1954). In an Alaskan sample of 61 Pacific common eiders, mollusks constituted 46 percent, crustaceans 30.7 percent, and echinoderms 14.4 percent of the total food volume found. A sample of 96 American common eiders showed the same relative importance of these three food sources but suggested a higher total consumption (81.7 percent) of mollusks (Cottam, 1939).

In a six-study comparison of foods found in the digestive tracts of common eiders, mollusks, crustaceans, and echinoderms were found in all six, insects and anemones each in three, miscellaneous animals in two, and polychaetes, fish and fish eggs, jellyfish, and plants/algae in one each (Goudie, Robertson and Reed, 2000).

Cottam (1939) described the usual foraging behavior of common eiders as diving from a point usually just beyond the surf, detaching mussels from rocky bottoms, and taking relatively few species of animal foods each meal. The birds dive to moderate depths, forage particularly at low tide, and at least during fall and winter apparently feed only during daylight. At night they move to the open ocean, sometimes many miles from their foraging areas.

Sociality, densities, territoriality. The high degree of sociality exhibited by nesting eiders is well established and perhaps reflects their island-nesting tendencies plus the gradual buildup of nesting groups in local areas protected from avian and mammalian predators. Thus Choate (1967) reported overall nesting densities of 3.8 to 8.9 nests per 1,000 square feet on various islands in Penobscot Bay. On smaller study areas within these islands, higher densities (4.3–136.4 nests per 1,000 square feet) were found. However, he found no relationship between nesting densities and nesting success.

Guignion (1969) reported an even higher average nesting density (16 nests per 1,000 square feet) on one islet that he studied. These densities are apparently well above those found by Cooch (1965), who noted that one ridge 8 acres in area supported more than 100 nests (or about 0.3 per 1,000 square feet), or Marshall (1967), who reported up to 100 breeding birds per acre (or about 2 per 1,000 square feet).

Manning et al. (1956) noted that there were an estimated 250 Pacific common eider nests on a sandy, sparsely vegetated island measuring 150 × 70 yards, or a maximum of about 2 acres. Thus it would appear that nesting "territories" of 100 to 300 square feet are not uncommon in dense eider colonies. Prior to the start of incubation the pair might spend a good deal of time resting on communal loafing areas, but Cooch (1965) did not observe the male actively defending his mate on such areas. Since males also only visit the nest site when the female is in the process of egg-laying, it is apparent that there can be no effective male defense of the nest site either.

Interspecific relationships. The obviously close relationship of the common and king eiders would suggest possible competition for food, nests, or other aspects of their biology. Pettingill (1959, 1962) observed mixed pairing and reported a presumed hybrid that apparently resulted from male king eiders mating with female common eiders nesting in Iceland, which is outside the king eider's breeding range. In areas where both species nest, their differences in preferred nesting habitats and substrates would probably tend to reduce such contacts.

The preferred foods of common eiders, such as *Mytilus* mussels and crabs, are virtually identical to those of king eiders, scoters, and, to a lesser extent, some of the other sea ducks (Cottam, 1939). However, the usual abundance of such foods makes it unlikely that significant competition normally occurs. Perhaps the most important relationships with other birds are those with ravens and large gulls, such as the great black-backed and glaucous. Nearly all nesting studies have indicated considerable losses of eggs and young ducklings to gulls, as noted earlier. Diseases and parasitic infections have also been reported as possible causes of juvenile mortality by some investigators.

General activity patterns and movements. That eiders exhibit marked daily periodicities in their behavior was established by Gorman (1970), who determined the frequencies of male displays throughout the daylight hours. He determined a dawn and dusk peak of display, with the one associated with sunrise higher than the one at sunset, and several bursts of activity throughout the day, interspersed with resting periods. There was also a tidal periodicity, with display being higher during periods of flood and ebb tides and lower during times of high and low tides, when eiders are roosting and foraging, respectively.

There is considerable variation in the local and migratory movements of common eiders. Atkinson-Willes (1963) considered the eider population in Great Britain to be entirely sedentary, and perhaps the same applies to the birds breeding in New England. The Hudson Bay population might also be fairly sedentary, but this cannot be true of the northern common eider or the Pacific common eider populations. Mossbech et al. (2006) found that although common eiders nesting on the Yukon-Kuskokwim Delta were essentially nonmigratory, wintering no more than 400 kilometers from their nesting sites, others banded in Arctic Canada and western Greenland have migrated in excess of 2,000 kilometers. Still others nesting along the coast of the Beaufort Sea migrated through the Bering Strait to winter on polynyas or among pack ice up to 1,600 kilometers away.

Cooch noted that the order of fall migration of the northern common eider from Cape Dorset was the reverse of that in spring, with the males and subadults of both sexes apparently leaving first, followed by adult females and their offspring. In spring the males are first to arrive.

Social and Sexual Behavior

Flocking behavior. The flock sizes of migrant birds arriving at Cape Dorset in spring was studied by Cooch (1965), who noted that the earliest flocks consisted of about 10 to 17 birds, but flock sizes progressively diminished as the sex ratio equalized, so that the latest arrivals were in groups of about 2 to 4 birds, with the sexes equally represented. During the fall migration, flock sizes are considerably larger. The data of Thompson and Person (1963) for Point Barrow illustrate this nicely. Between mid-July and early September they estimated that a million eiders (king and Pacific common) fly over this point on the way to molting and wintering areas. The usual sequence for both species is for adult males to arrive first, followed by flocks of mixed sexes, and lastly juveniles. The mean flock size for both species was 105 birds, and the modal flock size category was 26 to 50 birds. The largest flock seen was estimated at 1,100 birds.

Pair-forming behavior. In captive birds pair-forming behavior in common eiders begins in winter shortly after the birds have attained their nuptial plumage, and it is probable that some winter courting activities also occur in wild birds (Hoogerheide, 1950). McKinney (1961) thought that pair formation in European common eiders might occur at any time from March to May. Kenyon (1961) reported May pairing in the Pacific common eider, and Cooch (1965) reported a high incidence of courtship behavior during late May.

The pair-forming patterns of the European and Pacific races of the common eider were described and compared by McKinney (1961), and limited observations on the northern and American races indicate that these races have male display patterns nearly identical to those of the European race (Johnsgard, 1965; Cooch, 1965). Male pair-forming displays of common eiders consist of a variety of variably ritualized comfort movements such as preening, bathing, wing-flapping, and wing-shaking, and several different and more ritualized "cooing movements" associated with neck inflation and dovelike calls. McKinney recognized three individual types of cooing movements, as well as two compound combinations of them that he observed in the European but not in the Pacific race. Although a turning-of-the-back-of-the-head toward the female

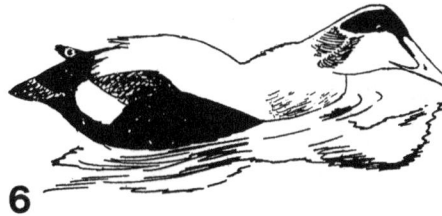

Displays of the common eider, including (1) female inciting, (2) male chin-lifting, (3–4) male bill-tossing, (5–7) male reaching sequence, and (8) male precopulatory wing-flapping.

does not occur, a conspicuous lateral swinging movement of the head is present ("head-turning") and often precedes or follows other displays.

McKinney believed that the overall function of social courtship in eiders is to allow individual pair formation to occur, but he did not observe specific instances of mate selection by females. He suggested that the more aggressive male that swims closest to the female might tend to intimidate other males, and, if he is accepted by the female, is effective in keeping other males away from her.

Copulatory behavior. McKinney (1961) observed copulations in the European common eider from late February until early May, or more than two months before the first eggs were laid. The female assumes a prone posture early in the precopulatory situation, although it is at least as often true that the male initiates the copulatory sequence. His displays include virtually all of those that occur in social courtship situations but include relatively few cooing movements. Instead there is a high incidence (in the European race) of preening, bathing, neck-stretching, and shaking. The order of these displays is not rigid, but mounting is usually immediately preceded by head-turning or a cooing movement.

In the Pacific race, bill-dipping, bathing, preening, and shaking are the most common precopulatory displays, and the display immediately prior to mounting is usually shaking or head-turning. During treading the male holds the female's nape, and in both races that were studied the male then typically performs a single cooing movement display and swims away from the female while head-turning. The female's postcopulatory behavior is variable but usually includes bathing (McKinney, 1961).

Nesting and brooding behavior. Cooch's (1965) study of the northern common eider at Cape Dorset provides a useful summary of eider nesting behavior. Female eiders evidently often return to nest sites used in prior years and prepare them for reuse by churning up the old detritus with the bill to permit air to circulate and dry out the site. New sites are usually prepared on the same day the first egg is laid. Most females visit their nests only at high tide during the egg-laying period and might begin to deposit down with the first egg or later. Most females begin incubation after laying their third egg, according to Cooch, even though a fourth might be laid. They might drink and bathe during the early part of incubation, but evidently little or no food is consumed during the entire period prior to hatching.

The first egg laid is the first to hatch, and up to an additional 24 hours might be required before all of the eggs have hatched. One or two additional days might be needed to dry the young thoroughly and prepare them to leave the nest. The brood is then led to tidal pools, sometimes as far as 1,000 feet from the nest. At first the ducklings feed almost entirely on the surface but gradually gain in diving efficiency. They begin to feed on mosquito larvae a few days after hatching and later shift to other invertebrate foods. As they develop, a tendency for brood merger becomes increasingly evident, and large crèches of eider ducklings typically form.

Studies by Gorman and Milne (1972) on crèche behavior of common eiders in Scotland indicate that the adult females guarding crèches were mainly birds that had recently hatched young and that they remained with the crèche only a few days before leaving it, presumably to forage and recover the body weight lost during incubation. Crèche behavior is thus not typical of eider populations in areas where food sources suitable for both adults and ducklings are present in the same habitats.

Postbreeding behavior. Males typically desert their mates when incubation starts, or very early in the incubation period. The males then move back out to sea and probably begin the postbreeding migration to molting areas. Cooch believed that females molt while their broods are still flightless and that both the young and the females attain flight at about the same time. However, he noted only a few flightless adults in his study area, and it is probable that at least some of the females also undertook a molt migration to other areas.

Molt migrations are typical of most common eiders except for the Hudson Bay population, which molts along offshore islands within its breeding range. American common eiders molt in a variety of areas, including the northern shores of the St. Lawrence estuary and the Gulf of St. Lawrence, off southwestern Nova Scotia, and in Maine within the Petit Manan archipelago and off Metinic Island (Baldassarre, 2014). Northern common eiders are known to molt in various areas, including along the northeast and southeast coasts of Baffin Island (Abraham and Finney, 1986) and in Ungava Bay (Nakashima, 1986).

As noted earlier, large numbers of Pacific common eiders annually fly past Point Barrow in late summer to molting areas along the Beaufort Sea coast. Some of those birds that bred near Bathurst Inlet continue on to winter off the southeastern coast of the Chukotsk Peninsula, or off the Russian coast (Baldassarre, 2014).

King Eider
Somateria spectabilis (Linnaeus) 1758

Other vernacular names. None in general use.

Range. Breeds in a circumpolar distribution in northeastern Russia, St. Lawrence Island, northern Alaska, and the Arctic coast of Canada including most of Canada's Arctic islands north to northern Ellesmere Island, south to the northwestern coast of Hudson Bay and northern Quebec. Winters in the northern Pacific and Bering Sea, especially along the Aleutian Islands, sometimes south to California; also winters on the Atlantic coast from southern Greenland to Newfoundland and along the New England coast, occasionally to Georgia and sometimes far inland.

Subspecies. None recognized.

Measurements. *Folded wing:* Delacour (1959): Males 275–290 mm, females 260–282 mm. Palmer (1957): Males 271–292, ave. (12) 280 mm, females 251–282 mm, ave. (11) 270 mm.
 Culmen: Delacour (1959): Males 28–34 mm, females 30–35 mm. Palmer (1957): Males (dorsal midline from edge of feathers) 26–34 mm, ave. (12) 30.7 mm, females 21–36 mm, ave. (11) 33.7 mm.

Weights (mass). Nelson and Martin (1953): 14 males ave. 4.0 lb. (1,814 g), max. 4.4 lb. (1,995 g); 9 females ave. 3.6 lb. (1,633 g), max. 4.1 lb. (1,859 g). Thompson and Person (1963): 41 adult males ave. 3.68 lb. (1,668 g); 140 adult females ave. 3.46 lb. (1,567 g).

Identification

In the hand. Easily recognized as an eider on the basis of its sickle-shaped tertials and the extension of feathering along the sides and top of the bill, the king eider is the only eider (see also surf scoter) in which the feathering on the culmen extends farther forward than the lateral extension near the base of the bill. The unfeathered area between these two extensions is generally wider than in common eiders, particularly in males, where it is greatly enlarged in adult males. Females are the only large eiders (folded wing 260–282 mm) that exhibit crescent-shaped dark markings on the mantle and sides of the body.

In the field. On the water, male king eiders show more black color than any of the other eiders, with the rear half of the body appearing entirely dark except for a narrow white line where the wings insert in the flanks and a white patch on the sides of the rump. The black "thorn feathers" among the rear scapulars protrude above the back conspicuously; in the common eider these either are not evident or are white (Pacific race). The enlarged reddish base of the bill is evident at great distances, even when the birds are in flight. Females are distinctly more reddish than female common eiders; they have crescent-like body markings

The North American breeding (hatched) and wintering (shaded) ranges of the king eider.

and a definite decumbent crest, which corresponds to the unique bluish feather area on the male. In flight, king eiders are slightly less bulky and ponderous than common eiders, and in a flock containing males the discontinuity of the white on their breasts and upper wing-coverts caused by the black back color is plainly evident. Calls of the female king eider include loud *gog-gag-gog* notes, like the noise produced by a hammer hitting a hollow wooden wall, and males utter tremulous cooing sounds during courtship.

Age and Sex Criteria

Sex determination. After the loss of the juvenal plumage, males exhibit some white on the breast or back, and even while in postbreeding ("eclipse" or basic) plumage, they retain some grayish or white feathers among the upper wing-coverts.

Age determination. Young females are probably not readily separable from adults after they lose their juvenal notched tail feathers. First-year males have a generally limited amount of white on the breast and rump; in second-year males the median wing-coverts are margined or dusky, and in older males these feathers are entirely white.

Distribution and Habitat

Breeding distribution and habitat. The North American breeding distribution of the king eider is not quite as extensive as that of the common eider, and in general it is more typically Arctic-oriented, with the southernmost breeding occurring at about 55°N latitude, but mostly nesting above 65°N latitude. In Alaska the king eider apparently breeds mainly along the Arctic Coastal Plain, mostly east of Barrow. To the west it breeds sparingly to Point Hope, Tigara, and Cape Thompson; to the east it breeds fairly commonly to Demarcation Point; an average of about 15,000 birds were found during 1992–2008 surveys (Larned, Stehn, and Platte, 2009).

The Canadian breeding range is extensive, occurring along the Arctic coastlines of Yukon Territory, Northwest Territories, coastal Nunavut, Southampton Island, and south on the western coast of Hudson Bay to southeastern Nunavut, and rarely or locally as far south as Cape Henrietta Maria and South Twin Island. There are a few known areas of breeding records from the Arctic coast of Quebec, but Labrador breeding is apparently unproven. Breeding also occurs on most of the high-Arctic islands in the Franklin District, northward to northern Ellesmere Island and east to western and northeastern Greenland.

The breeding population of Victoria Island was estimated at 800,000 birds in the 1960s (Parmelee et al., 1967), but by the late 1990s was down to 150,000 (Dickson et al., 1997), and to about 25,000 in 2005 (Conant, Roetker, and Groves, 2006). There were similar declines at Banks Island (from 100,000 to 40,000) from the 1960s to the 1990s, when an estimated 19,400 were present in the Queen Maud Gulf Bird Sanctuary (Dickson et al., 1997).

The preferred breeding habitat consists of freshwater ponds on Arctic tundra or amid lakes and streams not far from the coast. In a few instances they have been found nesting just above the high tide lines of seacoasts, but more commonly they might be found in the vicinity of fresh water (Godfrey, 1986; Snyder, 1957).

King eider, pair in breeding plumage

Wintering distribution and habitat. Wintering in Alaska occurs on at least the eastern Aleutian Islands east to Kodiak Island and the adjacent coast of the Alaska Peninsula (Gabrielson and Lincoln, 1959). The birds also winter in large numbers around St. Lawrence Island, making up the majority of the 50,000 or so eiders that occurred there in the late 1950s (Fay, 1961).

The population of eastern North America mainly winters from southern Greenland to Newfoundland, in the Gulf of St. Lawrence, and along the Maritime Provinces, with smaller numbers reaching the New England states. According to the records of king eiders seen in the New England states during the winter of 1970–71, most of the flocks occurring that far south contain 10 to 20 birds and consist of females and immature males.

The dividing line between those king eiders wintering in the Bering Sea and those that move eastward toward Greenland and the North Atlantic is not known, but some individuals from as far west as King William and Southampton Islands have been found during later summer off western Greenland (Godfrey, 1986). Parmelee et al. (1967) judged that part of the population breeding on Victoria Island probably migrated in part to the west and partly to the east.

Wintering habitats consist of the open sea or coastlines that have sources of food (mussels, etc.) at depths sufficiently shallow to permit easy diving. The birds tend to forage farther from shore than do long-tailed ducks and scoters, although they are perhaps less-efficient divers than long-tailed ducks. Cottam (1939) summarized evidence favoring the view that king eiders forage in deeper waters than do common eiders, and indeed deeper than any other duck with the possible exception of the long-tailed duck. There is one record of a bird apparently diving to a depth of 180 feet and returning with mollusks in its gullet.

General Biology

Age at maturity. There is little evidence for determining age at maturity in king eiders. All the known-age female eiders nesting at Karrak Lake, Nunavut, were at least three years old (Mehl, 2004). On the basis of plumage succession, Bent (1925) and Dementiev and Gladkov (1967) judged that maturity is probably reached during the third year of life.

Pair-bond pattern. Pair-bonds are renewed yearly during social courtship. This process has not been studied in wild birds, but at least in captivity the period of courtship display occurs over several winter and spring months.

Nest location. Manning et al. (1956) noted that on Banks Island the king eiders usually nested beside lakes, on small islands in lakes, or in low marshy country, but they sometimes utilized almost bare hillsides. Siberian observers report nesting on low mossy tundra near small lakes or rivers at varying distances from the sea, on dry grassy tundra, and occasionally also on high-growing tundra with knotweed (*Polygonum*) present.

Nests are usually well scattered, but where predation by foxes is prevalent dense nesting groups sometimes occur on river islands (Dementiev and Gladkov, 1967). Parmelee et al. (1967) found about 25 nests on dry, often rocky, slopes on Victoria Island, none of which were near water and one of which was about a quarter mile from water. The most closely spaced nests they found were 200 yards apart.

Clutch size. Parmelee et al. (1967) reported clutch sizes for 27 nests, which averaged 5.04 eggs. A similar average clutch (5.4 eggs) was found at Karrak Lake for 37 nests (Kellett and Alisaukas, 1997). The normal clutch range appears to be 3 to 6 eggs, although larger, very probably multiple, clutches of up to 16 eggs have been reported (Bailey, 1948).

Incubation period. Parmelee et al. (1967) found that the incubation period of naturally incubated eggs was 22 to 24 days. This is close to the 22 to 23 days reported for artificially incubated eggs (Johnstone, 1961, 1970).

Fledging period. Fledging occurs at about 50 to 60 days (Kear 2005).

Nest and egg losses. Several recent studies on nesting losses have reported on egg and nest success. Studies by Karrett and Alisaukas (2000) at Karrak and Adventure Lakes on about 1,000 nests resulted in varied nest success rates ranging from 30 to 89 percent, with the highest success rates on small islands and

King eider, males and female

King eider, male in flight

on well-isolated larger islands. There predation caused an estimated 66 percent of nest failures, with Arctic foxes, glaucous gulls, herring gulls, and jaegers evidently being the most significant predation factors.

On a large sample of 289 nests on the Alaskan North Slope, nesting success ranged from 21 to 57 percent (Bentzen et al., 2008). Predators were similarly mostly large gulls, Arctic foxes, and ravens. Earlier writers have also reported egg losses to both foxes and gulls (Phillips, 1926).

Juvenile mortality. No doubt gulls and jaegers consume some newly hatched king eider ducklings; the large "nurseries" of ducklings of both this species and the common eider have usually been regarded as a means of reducing the magnitude of such losses. Mehl and Alisauskas (2007) estimated that among 34 broods at Karrak Lake, 23.5 percent were supplemented by brood amalgamation, and Parmelee et al. (1967) noted that such amalgamated crèches might total up to more than 100 ducklings being tended by as many as 9 females.

Most brood mortality occurs during the first few days following hatching, Among 100 ducklings, apparent brood survival over the first 2 weeks after hatching was 31 percent for broods and 10 percent for individual ducklings. Late-hatching broods suffered higher mortality than earlier-hatching ones, with gulls and Arctic foxes being the most significant predators of ducklings (Lamothe, 1973; Mehl and Alisauskas, 2007).

Estimates of mortality for fledged juveniles are rarely available for sea ducks, but Oppel and Powell (2010) were able to use satellite telemetry to estimate first-year mortality rates. They calculated a 67 percent first-year survival for 49 birds, and a second-year survival of 100 percent for 21 birds. Among the 2-year-old females, 8 of 9 returned to their natal area in their second year, but none of 3 males returned.

Adult mortality. At least two estimates of adult mortality rates are available. Among 147 adult females banded at Karrak Lake from 2001 to 2003, the annual mortality rate varied from 60 to 91 percent (Weyland et al., 2008b), and over a six-year period (1996–2002) averaged 87 percent (Mehl, 2004). The two oldest known ages attained by wild birds were in excess of 10 and 15 years (Salomonsen, 1965; Suydam, 2000).

General Ecology

Food and foraging. Foods much like those taken by the common eider appear to make up the diet of adult king eiders, with an emphasis on bivalve mollusks (especially *Mytilis* mussels), crabs (especially *Cancer* and *Dermaturus*), and echinoderms (especially sand dollars and sea urchins). Probably no other duck consumes such a high incidence of echinoderms as the king eider, nor are such a wide variety of echinoderm types usually consumed. Sand dollars and sea urchins are favored foods, but sea stars, brittle stars, and sea cucumbers have also been found in king eider digestive tracts (Cottam, 1939).

Evidently eelgrass (*Zostera*) is one of the few plant foods of notable importance to king eiders, although relatively few specimens collected on the summer nesting grounds have yet been analyzed. Because the king eider forages so far from shore, in even deeper water than related species, virtually nothing can be said of its foraging periodicities or behavior.

Sociality, densities, territoriality. Apparently king eiders normally are not social nesters, and only in areas where small river islands provide protection from Arctic foxes do dense nesting colonies develop (Dementiev and Gladkov, 1967). More typically the nests are well scattered and several hundred yards from others of the species (Parmelee et al., 1967). Breeding densities are thus probably rather low in most areas, but no detailed estimates are available.

If the mid-1900s estimate of 800,000 king eiders mentioned earlier for Victoria Island was at all close to correct, the density then would have been at least 10 birds per square mile for the island as a whole, and much of its interior is probably unsuited for eiders. Banks Island, with a total land area of about 25,000 square miles, had a 1950s estimated king eider population of some 150,000 birds, or about 6 per square mile for the island as a whole (Manning et al., 1956).

Interspecific relationships. King eiders do not normally nest among common eiders, but in a few instances male king eiders have been seen intruding in common eider colonies. This has led to some instances of mixed pairing and possible hybridization (Pettingill, 1959, 1962).

The relationship of king eiders to gulls, ravens, foxes, and other possible predators of eggs and young is not yet established but is probably comparable to that indicated for the common eider. Parmelee et al. (1967) noted probable egg losses to jaegers and losses of both ducklings and adults caused by the native Inuit.

General activity patterns and movements. At least during the summer months, this high-Arctic species is probably active at all hours. At Point Barrow spring migration predominantly occurs during daylight, with a midday pause and with midmorning and midafternoon peaks. In late summer the return migration is more continuous, starting about daybreak (or 3:00 to 4:00 a.m.) and continuing virtually without interruption until sunset, or about 9:00 p.m. (Bent, 1925).

Social and Sexual Behavior

Flocking behavior. Mass assemblages of immature and molting birds are known to occur in Russia around Kolguev Island, and to a lesser extent at Vaigach Island and the west coast of Novaya Zemlya. The vicinity of Kolguev Island (Russia) is a major molting area for adults, and immatures remain there throughout the year in the tens of thousands.

Another area of congregation of immatures is along the coast of Russia's Chukotsk Peninsula and coastal Alaska from Point Barrow to the vicinity of Nunivak Island (Dementiev and Gladkov, 1967). In eastern North America the waters off the coast of west-central Greenland likewise attract vast numbers of molting birds (Salomonsen, 1968).

Pair-forming behavior. The distinctive pair-forming displays of this species have been described by Johnsgard (1964a, 1965) and Sherman (1965). The male displays include several ritualized comfort movements (bathing, wing-flapping, head-rolling, and general body-shaking, or upward-stretch) as well as two displays that are obvious homologues of the common eider's "cooing movements" and head-turning. Some of these displays have a remarkable uniformity in time-duration characteristics of the displays themselves as well as the intervals between displays occurring in sequence.

Sherman (1965) reported a strong tendency for successional "linkage" between certain displays, such as an association between wing-flapping and the upward-stretch, the upward-stretch and pushing, and bathing and wing-flapping. Wing-flapping is more highly stereotyped in the king eider than in any of the other eider species and conspicuously exhibits the male's underparts and throat markings during its performance (Johnsgard, 1964a, 1965).

Female pair-forming activities are virtually identical to those of common eiders, although the vocalizations produced are slightly different in the two species. Inciting appears to form a fundamental feature of social display and seems to be a primary means by which associations between individual males and females is achieved (Johnsgard, 1965).

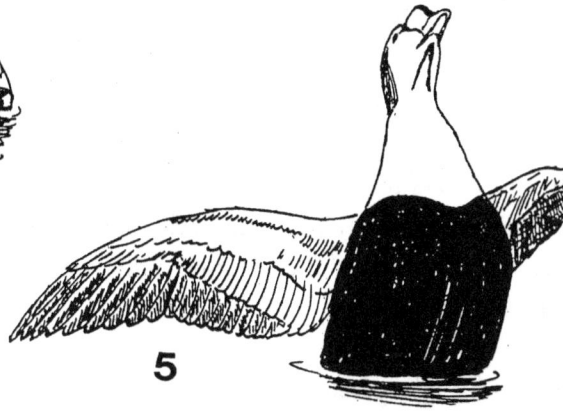

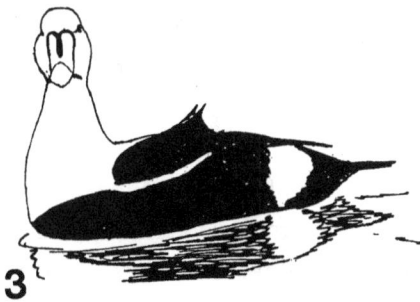

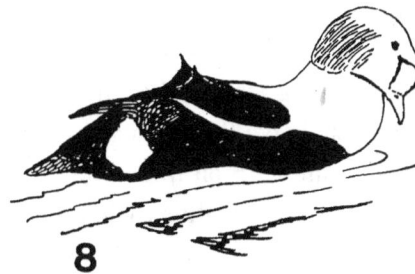

Male displays of the common eider (1–2) and king eider (3–8), including male (1) precopulatory preening, (2) precopulatory upward-stretch, (3–4) head-turning, (5) wing-flapping, and (6–8) reaching sequence.

Copulatory behavior. Like the situation in the common eider, female king eiders indicate a readiness for copulation by gradually assuming a prone posture as the male performs a nearly continuous series of displays, including the two major courtship postures (pushing and reaching) and, more typically, the four ritualized comfort movements. Of these, bathing occurs most frequently, followed in sequence by the upward-stretch, head-rolling, and wing-flapping. Bill-dipping and preening dorsally have also been seen.

In at least two of four cases the display performed just prior to mounting was wing-flapping, and in a third it was the upward-stretch. After each completed copulation the male released the female, performed a single reaching display, and then swam rapidly away from her while performing lateral head-turning movements (Johnsgard, 1964a).

Nesting and brooding behavior. Parmelee et al. (1967) have provided the most recent and most complete observations of nesting behavior by wild king eiders. During the egg-laying period the male closely attends his mate and follows her to the nest site for her egg-laying visits. Eggs are apparently laid at the rate of one per day. However, shortly after incubation begins the males desert their mates and rapidly move toward the coast to begin their molt migration. As with the common eider, it is likely that the female spends very little time off the nest during the 22- to 24-day incubation period. In at least one case, all of 6 eggs in one nest hatched within a 24-hour period.

Brood merging (aggregation behavior) is extremely common in this species and begins shortly after hatching. The number of females attending such crèches ("nurseries") varies, but up to 100 or more young have been seen together, with up to 9 females in attendance. Apparently many of the females that are displaced from their broods flock together and migrate out of the region before molting. The remaining females continue to attend the growing ducklings and might remain in the breeding areas until as late as September.

Postbreeding behavior. Salomonsen (1968) reported that at the peak of the late-summer molt migration to western Greenland, some 100,000 king eiders congregate in the waters off western Greenland. This total includes immature birds, some of which move directly into the area from their wintering quarters. Other immatures might approach the breeding range but fail to complete their migration and return to the molting area in June. Few adult females move to Greenland to molt but instead perform a later, shorter migration, probably to the vicinity of Clyde Inlet, Baffin Island. In September and October this population moves southward toward the ice-free areas of southwestern Greenland and on to Labrador and Newfoundland.

In the west, probably most birds fly west from breeding grounds in western Canada to the Beaufort Sea and on to molting areas in the Chukchi and Bering Seas, although some birds molt in scattered locations along the Chukotka Peninsula and western Alaska (Dickson, Suydam, and Balogh, 1999; Baldassarre, 2014). The late summer molt migration of king eiders past Barrow, Alaska, is justifiably famous and has been mentioned by several writers. Woodby and Divorky (1987) reported that an estimated 360,000 king eiders passed Point Barrow in one 10-hour period during spring migration.

Thompson and Person (1963) provided a good account of this migration. They reported that migrating eiders could be seen at almost any time over a 24-hour period, but counts made during morning and evening averaged about twice as high as those made during midday. They estimated that at least a million eiders flew past Barrow between mid-July and early September, including both common and king eiders. Reports of up to 75,000 king eiders per day crossing the Bering Strait in early to mid-May have also been made (Dementiev and Gladkov, 1967).

The eiders passing Barrow include many from the Alaskan North Slope, and 60 marked birds from this population spread out along the east coast of the Chukotsk Peninsula, with some reaching as far south as Kamchatka, as well as to St. Lawrence Island, and Alaska's Kvichak Bay (Philips et al., 2006). (See also the "Flocking Behavior" subsection at the beginning of this section.)

According to Phillips (1926), the king eider is less predictable than the common eider in its migratory behavior, with individuals more frequently appearing in large lakes in the interior parts of the continent than is the case with common eiders. However, most of the stragglers that appear during fall and winter on inland waters are immature birds (Bent, 1925).

Spectacled Eider
Somateria fischeri (Brandt) 1847

Other vernacular names. None in general use.

Range. Breeds in North America along the north coast of Alaska, from Wainwright east to about Camden Bay, and the coastal Yukon-Kuskokwim Delta, especially between Nelson Island and Hooper Bay. Also breeds in northeastern Russia on the Yena, Indigirka, and Kolyma deltas, and along the Bering Sea coast of Russia at Mechigmenskiy Bay. The only known wintering area of the world population is largely limited to small ice-free regions of the Bering Sea southwest of St. Lawrence Island.

Subspecies. None recognized.

Measurements. *Folded wing:* Delacour (1959): Males 255–267 mm, females 240–250 mm; Palmer (1976): Males 249–265 mm (ave. of 8, 256 mm); females 244–269 mm (ave. of 120, 252 mm).
   Culmen (from nostrils): Palmer (1976): Males 25–27 mm (ave. of 10, 26.2 mm); females 25–27.5 mm (ave. of 10, 26.4 mm).

Weights (mass). Nelson and Martin (1953): 8 males ave. 3.6 lb. (1,633 g), max. 3.8 lb. (1,723 g); 4 females ave. 3.6 lb. (1,633 g), max. 3.9 lb. (1,769 g). Lovvorn et al. (2003): 26 adult males (March) ave. 1,688 g; 12 adult females (March) ave. 1,550 g. Dau (1974): 48 adult males (May) ave. 1,458 g; 40 adult females (May) ave. 1,443 g.

Identification

In the hand. Spectacled eiders are easily recognized by the distinctive "spectacles" around the eyes, and by the fact that the lateral surface of the bill from its base to a point above the nostrils is wholly covered with short, velvety feathers. Females have brownish bodies with darker bars on the mantle and sides as in common eiders, but their smaller body size (maximum folded wing length 250 mm) readily distinguishes them from that species if the head and bill characters cannot be examined.

In the field. Adult male spectacled eiders in breeding plumage are unmistakable in the field; the white eye-ring surrounded by green is visible for several hundred yards. Otherwise the top half of the bird appears white, and the bottom half is a dark silvery gray, including the lower breast. Females are generally tawny brown, with pale "spectacles" and a dark brown triangular area between the eye and the bill. When females are crouching on nests these dark brown cheek markings are highly conspicuous and often reveal the female's presence.

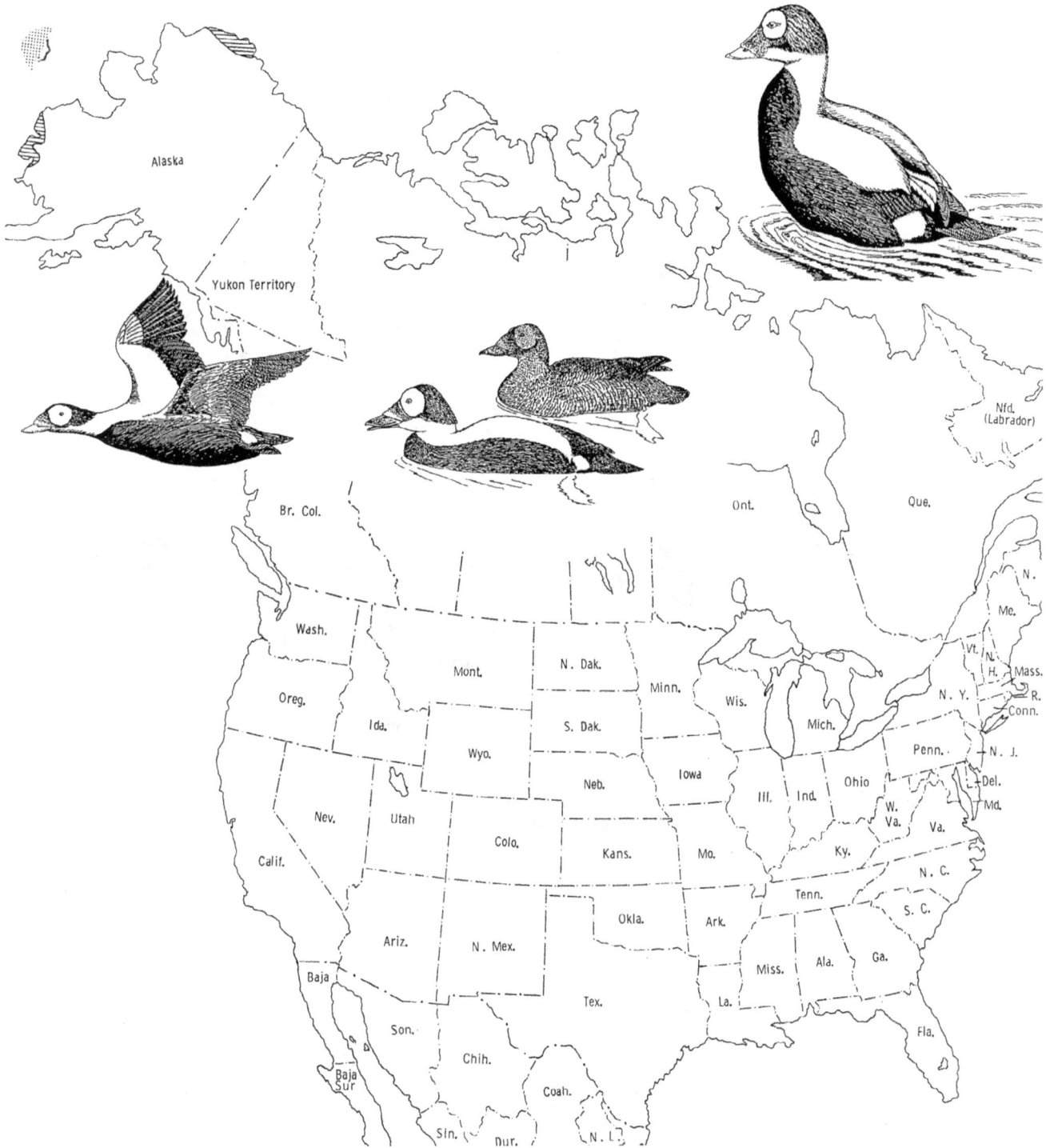

The North American breeding (hatched) and wintering (shaded) ranges of the spectacled eider. Regions of newly acquired or previously unrecognized breeding are shown by stippling.

Spectacled eider, male in flight

Spectacled eiders fly with considerable agility, and the extension of the blackish underparts to a point well in front of the leading edge of the wings serves to separate males from common eiders, while their white backs distinguish them from king eiders. Male spectacled eiders are unusually quiet, and their courtship calls are nearly inaudible beyond about 20 yards. Female vocalizations are very similar to those of the larger eiders.

Age and Sex Criteria

Sex determination. During the first fall and winter, the upperparts of young males are darker than those of females, and the underparts are faintly and duskily barred. Some white feathers appear on the neck and upperparts during the first winter. In second-year and older males the upper wing-coverts and tertials are pale grayish or white.

Age determination. First-year males have buffy-edged upper wing-coverts. The scapulars and probably also the upper coverts of second-year males are light gray, while those of older males are white. Juvenile females have spotted rather than barred underparts, and might still carry some notched tail feathers during the first fall, but somewhat older females are probably not distinguishable from adults on the basis of external features.

Distribution and Habitat

Breeding distribution and habitat. The known North American breeding distribution of the spectacled eider is mostly limited to a few areas along northern and western coastal Alaska. It breeds commonly on the Yukon–Kuskokwim Delta, especially in the vicinity of Igiak Bay and the adjacent coastal lowland tundra, north at least to the Askinuk Mountains. It breeds along the Arctic Coastal Plain from Wainwright east at least to about Camden Bay, perhaps to Demarcation Point. There have been occasional nestings on St. Lawrence Island (Fay, 1961). It also breeds in coastal tundra of northeastern Russia, mostly from Kromskaya Bay to Chaunskaya Bay, 136–162°E longitude (Dau and Kistchinski, 1977).

The preferred breeding habitat, judging from my observations of nesting sites at Igiak Bay, would seem to be rather luxuriant lowland tundra with small ponds and reasonable proximity to salt water. Fairly high grass of the past season's growth seemed to provide the basic nesting cover, and nearly all the nests that I found were placed fairly close to tundra ponds (Johnsgard, 1964b). Small lakelets in coastal tundra are also used for nesting in Siberian Russia (Dementiev and Gladkov, 1967).

Spectacled eider, male in breeding plumage

How far inland the birds ever move for nesting is still uncertain, but they evidently extend up the Kashunuk River some 25 miles to the vicinity of Chevak (Harris, 1966). Calvin Lensink (pers. comm.) has observed that, whereas Pacific and Steller's eiders are mostly limited to the coastal fringe, spectacled eiders often nest 5 to 10 miles up estuaries. Lensink estimated the Yukon–Kuskokwim Delta population at nearly 100,000 during the 1970s.

Wintering distribution and habitat. Almost the entire wintering area of this species, including the Asian breeders, is now known to occur in ice-free areas of the Bering Sea south of St. Lawrence Island, where more than 333,000 birds were first discovered during the winters of 1994 to 1997 (Petersen et al., 1999) The core wintering area, about 70 miles west-southwest of St. Lawrence Island, consists of about 570 square kilometers that encompass 84 percent of the total known wintering area (Petersen and Douglas, 2004). There is perhaps some additional wintering off the Russian coastline, south of St. Matthew Island and off Nunavut Island (Dau and Kistchinski, 1977). Although a few winter records exist for the Aleutian Islands (Murie, 1959; Gibson and Byrd, 2007), there is no indication that these islands are a major wintering area.

General Biology

Age at maturity. Age at maturity is not certain but generally thought to require two or three years (Bent, 1925; Dementiev and Gladkov, 1967). Males nearly two years old do exhibit male courtship behavior, suggesting that at least this sex is capable of breeding at the end of the second year of life, and their breeding plumage is only slightly different from that of third-year birds. About 50 percent of two-year-old females nest, and some females have been found to nest every year for as long as five consecutive years (Petersen, Grand, and Dau, 2000).

Pair-bond pattern. I have observed pair-forming behavior among wild birds during early June, and among captive birds during April (Pennsylvania) and May (England). It would thus seem that pairs are formed each spring and broken after the female begins incubation.

Nest location. According to our observations at Igiak Bay, nearly all of 13 nests our group found were placed within 3 to 4 feet of water, usually in dead marsh grasses surrounding small tundra ponds (Johnsgard, 1964b; Kessel et al., 1964). Of the 13 nests, the farthest from water were some 60 feet away, and the average of the remaining ones was 3.3 feet. Dau (1972) reported that nests are usually placed on sedge-dominated lowland areas, often on a shoreline (20 of 35 nests), a peninsula (8 of 35 nests), or on islands (7 of 35 nests). The greatest distance a nest was found from water was 240 feet, and the mean distance between neighboring nests was 389 feet.

Clutch size. Among 232 active nests observed in the Yukon-Kuskokwim Delta, the collective average clutch size was 4.57 eggs, with annual averages ranging from 4.0 to 5.2 (Calvin Lensink, pers. comm.). Other large samples from that area include a mean of 5.1 eggs (579 clutches) and 5.0 eggs (280 clutches), while

82 clutches from northern Alaska averaged 3.5 eggs (Petersen, Grand, and Dau, 2000). The eggs are probably deposited every other day (Brandt, 1943). Dau (1972) found no definite cases of renesting, but Michael Lubbock (pers. comm.) observed several apparent renests.

Incubation period. Dau (1974) stated that for three nests observed under natural conditions the incubation period lasted 24 days from the laying of the last egg. Flint and Grand (1999) reported a 22.3-day average incubation period for 19 nests.

Fledging period. Dau (1974) estimated that fledging by birds reared in the wild required about 50 days, with one marked individual attaining flight in no more than 53 days following hatching.

Nest and egg losses. Dau (1974) reported a nesting success of 90.9 percent for 33 nests in his study area, and a hatching success of 83 percent for 147 eggs in 29 nests. Hatching success (defined as a nest successfully hatching one or more eggs) varied from 14.8 to 66.8 percent for 579 clutches over 9 years on the Yukon–Kuskokwim Delta, and an overall 9-year mean of 41.4 percent. An overall 10-year average nesting success for 82 clutches around Prudhoe Bay averaged 26.9 percent (Petersen, Grand, and Dau, 2000).

Juvenile mortality. Brandt (1943) believed that fairly high losses of ducklings occurred shortly after hatching, judging from brood sizes (2–5) that he observed in the Hooper Bay–Igiak Bay area. Counts made there in 1950 (*United States Fish and Wildlife Service, Special Scientific Report: Wildlife, No.* 8) included a total of 33 spectacled eider broods, which averaged 5.2 young per brood. Between 1964 and 1971, 72 recently hatched (Class I) broods averaged 3.9 young per brood, according to Calvin Lensink (pers. comm.). Flint and Grand (1997) estimated a 34 percent duckling survival rate over the first 30 days for broods in the Yukon–Kuskokwim Delta; complete brood losses were highest during the first 10 days after hatching.

Brandt (1943), referring to Hooper Bay, believed that glaucous and glaucous-winged gulls were the primary predators of young ducklings, and observed a pomarine jaeger attempting to rob a nest. I found the parasitic and long-tailed jaegers more common than the pomarines, and on several occasions observed them swooping down on eider nests shortly after the female had been flushed from them.

Flint and Grand (1997) judged that survival of ducklings averaged 34 percent over the first 30 days, and that most duckling losses occurred during the first 10 days following hatching. By 30 days after hatching nearly half of all the marked females had lost their entire broods. Substantial losses (estimated at 28.6 percent) of ducklings later occurred between 30 days of age and fledging, with minks, Arctic foxes, and avian predators being the major predators.

Additionally, an estimated 15 percent of the brooding females were lost during the brooding period, so that 26–70 percent of the annual mortality in adult females occurred during that period (Flint and Grand, 1997). Grand et al. (1998) calculated that 54 percent of the annual mortality of successfully breeding females occurred between the time of hatching and when the females left the breeding area, about 56 days after their brood had hatched.

Spectacled eider, male and female

Adult mortality. Little information about adult mortality is available. Grand et al. (1998) reported that annual survival rates of adult females in Alaska was affected by lead exposure; those with elevated lead levels had annual survival rates of 44 percent versus 78 percent for those not exposed. One female spectacled eider was recaptured 18 years after having been banded as an after-hatching-year bird (Baldassarre, 2014).

General Ecology

Food and foraging. Petersen, Piatt, and Trust (1998) found that crabs and clams are apparently important winter foods, based on 38 birds that they analyzed. Cottam (1939) has summarized the little that is known of the foods of the spectacled eider, based on 16 adults collected between May and July. Animal foods, particularly mollusks, composed more than three-fourths of the food volume, while terrestrial and freshwater plant materials represented the remainder of the identified materials. Razor clams (*Siliqua*) constituted the majority of foods of 8 birds collected in May, and it seems probable that these and other bivalves are even more important on migration and wintering areas. Seemingly, the spectacled eider consumes a lower proportion of crustaceans than does the Steller's eider, and there were almost no echinoderm remains among the samples analyzed by Cottam.

An analysis of the stomachs of 5 juvenile spectacled eiders by Cottam indicated that insects made up the majority of food intake for such birds, with caddisfly larvae and their cases alone constituting more than a third of the total. A variety of other insects made up most of the remaining animal materials, whereas the seeds and plant fibers of mare's tail (*Hippurus*) were the largest single component of the plant materials found. Pondweeds (*Potamogeton*) and terrestrial crowberries (*Empetrum*) are also apparently important plant foods for immature birds.

Sociality, densities, territoriality. Since no large concentration of molting or wintering birds has ever been located, the degree of sociality during the nonbreeding season is unknown. In the nesting grounds, however, a surprising degree of sociality does seem to be present. My observations (Johnsgard, 1964b) suggested a slight clustering tendency for nests, although this was not obviously related to island-nesting or nesting in relatively secure locations. Brandt (1943) mentioned finding seven nests in a very restricted area, also suggestive of a colonial tendency.

Calvin Lensink (pers. comm.) found that in 1971 there were 23 nests per square mile in the Magak Flats behind Hooper Bay, which he considered to represent about half the normal density for the area. In a 231-acre study area of the lower Kashunuk River, from 8 to 34 spectacled eider nests were found per year in a three-year period, and the three-year average was 23, or 64 nests per square mile (*United States Fish and Wildlife Service, Special Scientific Report: Wildlife, No. 68*). I saw no evidence of a defended territory in my observations of nesting eiders at Igiak Bay, and frequently observed a breeding pair swimming within a few feet of another pair (Johnsgard, 1964b).

Interspecific relationships. Although Pacific common eiders have been seen in the company of spectacled eiders and have even been seen displaying among them (Johnsgard, 1964b), there is still no good evidence of notable interactions between the spectacled eider and any other species, although Pacific common eiders and Steller's eiders nest in the same general habitats as spectacled eiders. Probably the large gulls and the jaegers, and perhaps mammalian predators such as foxes and minks, are the most influential vertebrates in the breeding ecology of spectacled eiders.

General activity patterns and movements. During the long early summer days of June while we were at Igiak Bay, the spectacled eiders seemed to be active at all hours. Most display activities seemed to occur in early morning and late afternoon, while several copulations were seen near midday. Males and females that had completed their clutches or had lost their clutches early in incubation seemingly spent the entire day on one of the rivers, resting along the shore or diving for food in the river. Immature birds evidently largely remained at sea, since we saw only one obviously immature male, and Conover (1926) reported that, except for a single immature female, only fully adult birds were seen by their party.

Social and Sexual Behavior

Flocking behavior. Brandt (1943) noted that spring migrant arrivals in the Hooper Bay region in May were generally seen in groups of 10 to 30 birds with a maximum of 40. When I arrived in that area (early June) most of the spectacled eiders had initiated nests. Surplus adult males were still to be found on the rivers in groups of up to 14 birds, which seemed to be constantly on the watch for lone females (Johnsgard, 1964b).

Pair-forming behavior. Since most pairs had already been formed when I arrived at Igiak Bay in early June, it seems likely that the majority of the pair-forming behavior occurred while the birds were still in migration, or just after arrival at the breeding grounds. The pair-forming behavior I observed was primarily confined to unmated males whenever they encountered a lone female.

Pair-forming displays of this species were first described on the basis of my observations of these wild birds (Johnsgard, 1964a) and were later substantiated by observations of display in captives. Like the other eiders, females perform inciting movements and calls, which provide the basis for social courtship. The males respond with displays that include the usual ritualized comfort movements of eiders (wing-flapping, preening, bathing, shaking, and head-rolling), and two more specialized displays. One of these is a rapid backward rearing movement that exposes the blackish chest, or the same movement preceded by a preliminary movement of the head forward and slightly downward. This latter form approaches the reaching display of the king eider, whereas the former is clearly closer to the rearing display of the Steller's eider.

The male spectacled eider also performs a pushing display very much like that of the king eider, and a backward bill-toss followed (and sometimes also preceded) by a forward neck-jerk that is equivalent to the combination of two cooing movements of the common eider. In general, the spectacled eider has a display repertoire that merges elements of the genera *Polysticta* and *Somateria* and seems to provide a behavioral "link" between these types.

Copulatory behavior. Only four copulation sequences have been observed by me (Johnsgard, 1964a), but the general pattern seems to be much like that of other eiders. After the female assumes a prone posture, the male performs a nearly continuous sequence of movements; in order of observed decreasing frequency they are preening behind the wing, preening dorsally, pushing, bathing, head-rolling, bill-dipping, and wing-flapping. In at least three cases the male performed only a single shaking (upward-stretch) display, and always immediately prior to mounting. After treading, the male released the female, performed a single head-forward-rearing display, and then produced a few lateral head-turning movements while the female began to bathe.

Nesting and brooding behavior. Female spectacled eiders typically construct their nests in grassy flats on the islands or along the periphery of tundra ponds, making a slight depression that is scantily lined with grass stems. Invariably a substantial amount of down is present (Brandt, 1943). On the Yukon–Kuskokwim Delta, 41 percent of 208 successful nests were located on pond shorelines, 36 percent were on islands, and 23 percent were on peninsulas (Dau, 1974). On the Arctic Coastal Plain 39 percent of 50 nests were on

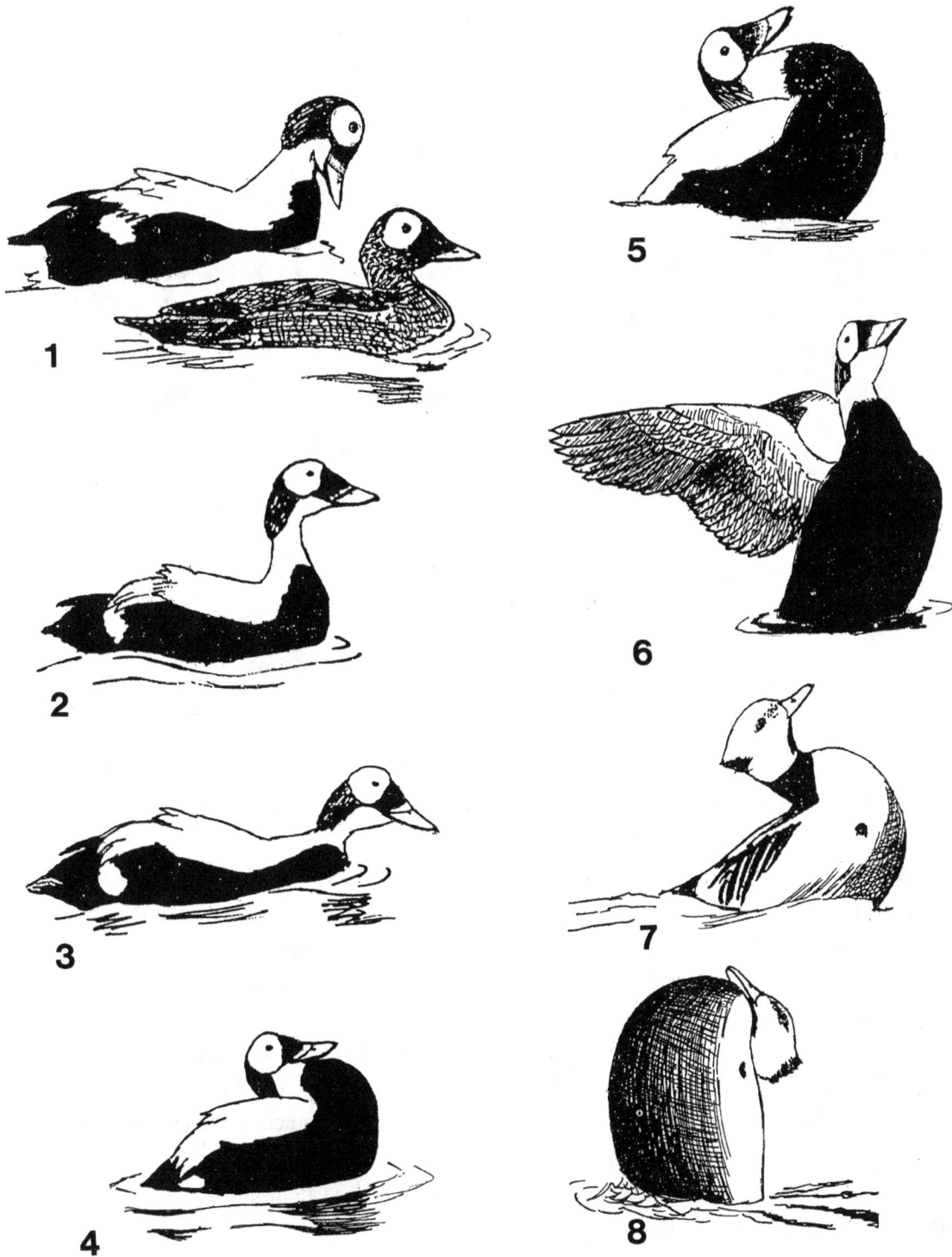

Male displays of the spectacled eider (1–6) and Steller's eider (7–8), including (1) neck-jerking, (2–4) head-forward–rearing sequence, (5) independent rearing, (6) wing-flapping, and (7 and 8) rearing (two views).

Spectacled eider, female incubating (Alaska)

islands, 24 percent on peninsulas, 21 percent on shorelines, and 14 percent on the rims of permafrost polygons (Bart and Earnst, 2005).

Once incubation begins the female is extremely reluctant to leave her clutch and can often be approached to within a few feet, if not actually touched. Males evidently remain in attendance until incubation is well underway; Brandt believed that they remained until the eggs are about to hatch. However, there is typically a mass exodus of males from the breeding areas late in June, when they fly back out to sea and presumably undertake a molt migration (Johnsgard, 1964b).

Relatively little is known of the brooding behavior, but individual families seemingly remain relatively intact and there is no early movement of young to open water. Dau (1972) stated that mixed broods are not common and that the young are reared to fledging on fresh to slightly brackish water areas that are probably within a mile or two of the nest. Crowberries (*Empetrum nigrum*) are preferred foods for both the female and the young, and some salmonberries (*Rubus arcticus*) are also eaten. Grasses and sedges, as well as some insects, are apparently consumed in considerable quantities.

Postbreeding behavior. The movement of adults and young are largely unknown once they leave their breeding grounds, but birds from both Alaskan and Russian breeding areas are known to molt at Russia's

Spectacled eider, male portrait

Mechigmenskiy Bay (170°W longitude), where 550,000 males and females were seen in September of 1994 (Petersen, Rand, and Dau, 2000). Movements away from mainland Alaska might occur in late July, August, or early September (Gabrielson and Lincoln, 1959). Williamson et al. (1966) reported seeing birds in the Cape Thompson area until as late as September 26. The only molting area thus far reported in Alaska is near Stuart Island, Norton Sound (Dau, 1972). In 1994 about 4,000 birds were seen there in early September (Petersen, Rand, and Dau, 2000).

Steller's Eider
Polysticta stelleri (Pallas) 1769

Other vernacular names. Soldier duck

Range. North American breeding is now mostly limited to Alaska's Arctic Coastal Plain, where the Steller's eider breeds primarily around Barrow, west to the vicinity of Wainwright, and east probably to about Prudhoe Bay. Previously a common breeder on the Yukon–Kuskokwim Delta, but very rare there by the mid-1990s. In North America, winters along the Aleutian Islands, the Kodiak Archipelago, and the Alaska Peninsula, rarely as far south as the Queen Charlotte Islands. Also breeds in Arctic Russia, mostly between the Yana and Indigirka River deltas, and east of the Lena River Delta.

Subspecies. None recognized.

Measurements. *Folded wing:* Delacour (1959): Males 209–217 mm, females 208–215 mm. Fredrickson (2001): Males 204–223 mm, ave. of 17, 212.2 mm; females 205–226.5 mm, ave. of 14, 211.5 mm.

Culmen: Delacour (1959): Males 36–40 mm, females 35–40 mm. Palmer (1976): Males 36–43 mm, ave. of 19, 39.7 mm; females 37–48 mm, ave. of 14, 43.8 mm.

Weights (mass). Fredrickson (2001): Males 719–969 g, ave. 887 g; females 780–945 g, ave. 852 g. Nelson and Martin (1953): 6 males ave. 1.9 lb. (861 g), max. 2.1 lb. (951 g); 6 females ave. 1.9 lb. (861 g), max. 2.0 lb. (907 g). Palmer (1976): 29 males ave. 759 g; 28 females ave. 826 g, Quackenbush et al. (2004): 17 breeding females ave. 711 g.

Identification

In the hand. Quite different from the larger eiders, the Steller's lacks feathering along the side and top of the bill, but it does possess sickle-shaped tertials. Unlike those of any other species of diving ducks, these tertials are iridescent blue on their outer webs, as are the secondaries. Other distinctive features are the narrow blackish bill, with soft marginal flaps near the tip, and a relatively long (up to 90 mm), pointed tail.

In the field. Because of their small size and agility, Steller's eiders are more likely to be confused with dabbling ducks than with other eiders. The male's cinnamon-colored sides and breast are visible for long distances, as are the mostly white head and scapulars. The black markings around the eye and the rounded black spot between the breast and the sides are also unique. Females are most easily identified by their association with males. Their size and uniformly dark brown color is reminiscent of abnormally dark female mallards, and in flight they also exhibit a mallard-like contrasting white underwing surface and

The North American breeding (hatched) and wintering (shaded) ranges of the Steller's eider.

two white wing bars. However, in taking off they run along the water like other diving ducks. The most conspicuous female call is a loud *qua-haaa'*, whereas males produce only soft growling notes that are not audible over long distances.

Age and Sex Criteria

Sex determination. Following their female-like juvenal plumage, immature male Steller's eiders can probably be distinguished from females as early as their first winter by the presence of some tawny chest feathers and a few white feathers on the head, mantle, or upper wing-coverts.

Age determination. After the juvenal tail feathers have been lost, immatures are probably difficult to age, but first-year males are presumably recognizable by the absence of white on the middle or lesser wing-coverts. The age of reproductive maturity in this species is uncertain but is probably two years for females. Immature males exhibit a dull blue rather than a bright blue speculum, and their secondaries are duskily tipped rather than white, whereas immature females have a dusky rather than dull blue speculum.

Distribution and Habitat

Breeding distribution and habitat. The North American breeding range of this species is even more restricted than that of the spectacled eider. Conover (1926) and Brandt (1943) found the Steller's eider nesting commonly in the vicinity of Igiak (Kokichek) and Hooper Bays. No nests were found there in 1964 (Johnsgard, 1964b; Kessel et al., 1964). However, it was later reported by Harris (1966) to be uncommon on the salt tundra flats near the mouth of the Kashunuk River, not far to the south of Hooper Bay. In 1991 Kertell judged that the species had disappeared from the Yukon–Kuskokwim Delta, but it was found to be a rare breeder there during the mid-1990s (Flint and Hertzog, 1999).

It evidently does not breed in the Norton Sound area, and Williamson et al. (1966) found no evidence of breeding near Cape Thompson. Eggs have been collected at Wainwright, and it is still known to be nesting fairly commonly around Barrow. From Barrow eastward there are egg records from Admiralty Bay and Pitt Point (Gabrielson and Lincoln, 1959), and other more recent dispersed breeding sites occur along the Arctic Coastal Plain (Quackenbush and Suydam, 1999; Quackenbush et al., 2002, 2004; Stehn and Platte, 2009). Reportedly there have been occasional nestings of this species on St. Lawrence Island (Fay, 1961), but most of the summering birds there are immatures.

The breeding habitat of the Steller's eider is lowland tundra closely adjacent to the coast. Brandt (1943) noted that this species nested closer to Hooper Bay than did either the Pacific common or spectacled eider, using the tidewater flats having small eminences near a body of water for nesting sites. In Siberia it typically nests in lacustrine basins on mossy tundra (Dementiev and Gladkov, 1967).

Steller's eider, pair in breeding plumage

Wintering distribution and habitat. The wintering distribution of the Steller's eider includes the vicinity of Kodiak Island (Larned and Zwiefelhofer, 2001), the south coast of the Alaska Peninsula (Laubham and Metzner, 1999), and the Aleutian Islands (Gabrielson and Lincoln, 1959). Jones (1965) reported that much of this wintering population, which at the time of his observations totaled about 200,000 birds at its peak, concentrated on Nelson Lagoon, Izembek Lagoon, and Becheven Bay. In many years these birds arrive prior to their postnuptial molt and undergo it there, while in others they might arrive as late as November. At Nelson Lagoon, which supports the largest populations, the birds feed in fairly shallow waters, apparently on crustaceans (McKinney, 1965; Peterson, 1980, 1981).

In Izembek Lagoon, a similarly shallow and muddy tidal lagoon used abundantly by molting birds, they forage around the extensive beds of eelgrass that almost choke the bay. When I was there in early September of 1961 we were able to net and band hundreds of flightless adults of both sexes. Although Izembek Bay is only a short distance overland from Cold Bay, that deeper, rockier bay is used little if at all by Steller's eiders, but it is commonly utilized by king eiders (McKinney, 1959).

General Biology

Age at maturity. Not well established. The adult male plumage is probably attained during the second fall of life (Bent, 1925), and sexual maturity probably occurs during the second or third year (Dementiev and Gladkov, 1967).

Pair-bond pattern. McKinney (1965) observed a high incidence of pair-forming behavior among wild Steller's eiders between March 31 and April 28, although many birds had already become paired by that time. Brandt (1943) believed that the male remains near his incubating mate until the clutch is about ready to hatch and then leaves the area. Some males might remain with the female for up to two weeks into incubation (Fredrickson, 2001).

Nest location. Brandt (1943) reported that small elevations near water on tidewater flats are the preferred nesting sites of the Steller's eider. Other nests have been reportedly built in flat, mossy tundra, with a single instance of a nest having been found among rocks (Bent, 1925).

Clutch size. Bent (1925) mentioned 5 clutches that ranged from 6 to 10 eggs and averaged 7.2 eggs. A. M. Bailey (pers. comm.) reported 7 clutches from Barrow's, Wainwright, and Cape Simpson that ranged from 3 (probably incomplete) to 7 eggs and averaged 6.1. Brandt (1943) noted one clutch of 7 eggs, three of 8, and one of 9. If all 17 of these clutches are considered, the modal clutch size would be 7 eggs and the average 7.2 eggs. On the Lena River Delta, Russia, 32 nests averaged 6.21 eggs, and 6 renestings averaged 3.9 eggs (Solovieva, 1997). Thirty clutches from the Western Foundation of Vertebrate Zoology averaged 7.7 eggs (Fredrickson, 2001). The average egg-laying interval is 23.1 hours (Quakenbush et al., 2004).

Incubation and fledging periods. The incubation period is 26 to 27 days (Solovieva, 1997). The fledging period is about 40 days (Obritschkewitsch et al., 2001), a surprisingly short period.

Nest and egg losses. At Barrow, nesting success was highly variable over five years, ranging from 14.6 to 71.3 percent, in correlation with lemming abundance, with higher success during years of higher numbers of lemming, an alternative food source for predators (Quakenbush and Suydam, 1999). Quakenbush et al., (2004) reported that egg success was also highly variable, ranging from 0 to 42 percent over five years of study, and nest success (hatching 1 or more eggs from a nest) ranged from 0 to 35 percent.

Nests placed fairly close (within 150 meters) to those of pomarine jaegers, or within 800 meters of a snowy owl nest, had higher nest success rates (25 to 28 percent) than did other nest locations. Phillips (1926) judged that glaucous, glaucous-winged, and perhaps mew gulls were probably serious egg predators, as well as jaegers and possibly snowy owls. Percy (in Bannerman, 1968) mentioned that dogs sometimes caused high losses to nesting waterfowl, including eiders, in Siberia.

Steller's eider, male in breeding plumage

Juvenile mortality. No specific information is available regarding juvenile mortality. Brandt (1943) noted that the sizes of half-grown broods usually numbered about 3 to 4 young, and he thus believed that the mortality rate for the ducklings must be fairly high. Of 13 broods monitored by Quakenbush et al. (2004) over two years, only a single duckling fledged on average, and during one year 6 of 7 broods were totally lost.

Adult mortality. Little information is available about adult mortality as well. Jones (1965) reported that 17 band returns had resulted from banding 833 adult eiders at Izembek Lagoon in 1961 and 1962. One of the birds banded there was later recovered at the age of 12 years, 3 months, in Russia. Flint et al. (2000) calculated an 89.9 percent annual survival for females of this population, and a 76 percent survival for males.

General Ecology

Food and foraging. Cottam's (1939) summary of Steller's eider foods is still a primary source of information on this topic. Among 66 eiders taken between May and July, crustaceans (primarily amphipods) and mollusks (primarily pelecypods) constituted more than 60 percent of the identified foods. Soft-bodied crustaceans such as amphipods, isopods, and barnacles appear to be a favorite food of the Steller's eider. Among the bivalve mollusks, a number of species of clams and mussels were eaten.

Insects of quiet tundra pools were also eaten in quantity, such as the larvae of midges, caddisflies, and other insects. Sand dollars and polychaete worms were also found in some stomachs and perhaps represent important foods for the birds while on salt water. Plant foods made up only about 13 percent of the stomach contents by volume. (McKinney, 1965) observed wild Steller's eiders feeding on an accumulation of dead shrimp in Nelson Lagoon during early spring. Two winter samples mentioned by Cottam had eaten little other than amphipods and univalve mollusks.

Fredrickson (2001) summarized food data from eight additional sources, including five from Alaska. Of these, Metzner's (1993) data were based on the largest sample size (138 birds from Izembek and Kinzarof Lagoons, plus Cold Bay). Half of the total volume of foods analyzed consisted of crustaceans, and crustaceans comprised almost half of total food volumes in two of the other studies. Still other samples from birds obtained during winter and summarized by Bauer and Glutz (1969) also suggest the importance of small univalve mollusks, amphipod crustaceans, and, to a lesser extent, bivalve mollusks and isopod crustaceans. The soft-edged bill of this species also supports the possibility that soft-bodied invertebrates probably represent the primary food supply.

Although Phillips (1926) reported that while at sea the birds tend to frequent the roughest, deepest, and rockiest coastlines, there is little current evidence to support this view. Rather, they seem to prefer shallow bays with muddy or sandy bottoms, such as Izembek and Nelson Lagoons (McKinney, 1965). In this respect they differ from the harlequin duck, which forages to a greater extent on hard-shelled crustaceans such as crabs, and among mollusks specializes in chitons, which are typical of rocky shores. When foraging, all the eiders in a foraging group commonly dive simultaneously, sometimes in very large flocks of about a thousand (McKinney, 1965).

Sociality, densities, territoriality. Although highly social on wintering areas, with flocks often numbering in the thousands of birds, the Steller's eider seemingly shows no tendency for social nesting. Nests are evidently not clustered, and overall nesting densities seem to be low, based on available information. On a 231-acre study area of the lower Kashunuk River in Alaska, there were 3 Steller's eider nests in 1951, 1 in 1961, and 5 in 1962 (*US Fish and Wildlife Service, Special Scientific Report: Wildlife, No. 68*). In 1963 only 1 nest was found on this area (Harris, 1966). Thus, an average yearly density of 2.5 nests was typical, or about 1 nest per 100 acres.

Interspecific relationships. Phillips (1926) noted that the most common reported association of Steller's eiders has been with king eiders, but they have also been observed with Pacific common eiders and harlequin ducks. He believed their major enemies to include glaucous and glaucous-winged gulls, jaegers, snowy owls, and perhaps also mew gulls. McKinney (1965) noted that both bald eagles and gyrfalcons are obviously feared by Steller's eiders, and he thought that these eiders might be more

Steller's eider, males and female

vulnerable than the larger species to such aerial predators, since the larger eiders mainly feed in deeper waters and farther from shore.

General activity patterns and movements. Percy (in Bannerman, 1958) reported that during June and July in Siberia, the daily movements of the Steller's eider seemed primarily influenced by wind and ice conditions rather than by time of day. McKinney (1965) observed birds feeding in areas exposed by the receding tide at Nelson Lagoon. A similar situation obtains at Izembek Lagoon in my own experience; the birds followed the rising tide as it encroached on the shallow bay, and as the tide retreated they gradually moved back out to deeper waters, where they rested.

Jones (1965) reported on some migratory movements of Steller's eiders banded as flightless adults at Izembek Bay/Lagoon during the early 1960s. All but one of the 17 nonlocal recoveries he had obtained at that time were in Russia at points as distant as the Lena River, some 3,200 kilometers distant. The remaining recovery came from Point Barrow, indicating that birds from both sides of the Bering Strait move to this region to undergo their molt, at least in some years. Far more birds appear at Izembek Bay for molting in some years than in others, suggesting that the distance the birds travel prior to undergoing their molt must vary from year to year.

Social and Sexual Behavior

Flocking behavior. The strong gregarious nature of this species during winter and on migration has been mentioned by a number of observers. McKinney (1965) noted 15 flocks in early April that ranged in size from about a thousand to several thousand birds. He reported on the unusual degree of behavioral synchrony of these tightly packed rafts, particularly in their foraging behavior. Evidently the vernacular name "soldier duck" refers to their notable trait of synchronized behavior. McKinney believed that many of the differences in the social behavior and displays of the Steller's eider as compared with the common eider could be attributed to its stronger social tendencies and greater readiness to fly.

Pair-forming behavior. The social behavior and pair-forming displays of this species have been described by Johnsgard (1964a) and McKinney (1965). Like the larger eiders, several comfort movements have been ritualized and incorporated into the display repertoire, including shaking (upward-stretch), preening the dorsal region, bathing, and head-rolling. Additionally, a lateral head-turning and two types of vertical head movements are present. One is a rapid and rather limited upward chin-lifting (head-tossing) movement, while the other is a much more extreme backward movement of the head and neck in a rapid "rearing" motion.

A very frequent sequence of behavior is for the male to perform a single shake, swim toward a female while performing lateral head-turning, perform a single rearing display when close to the female, and then swim rapidly away again while performing head-turning. Males also perform short display flights of a few feet, after which they alight near the female with a conspicuous splash.

Females enter into social display by their performance of strong inciting gestures and calls, which seem to be equivalent in function to inciting in the larger eider species. McKinney noted that aquatic display is often interspersed with flights of various lengths, with a maximum observed duration of about three minutes.

Copulatory behavior. I have observed two completed copulations (Johnsgard, 1964a), and McKinney (1965) reported on a much larger number of completed copulation sequences. We both noted that bill-dipping, dorsal-preening, and bathing movements were the three most typical male precopulatory displays and were performed in a relatively constant sequence. Thus, a preening display would usually alternate with either bill-dipping or bathing. McKinney also observed a few instances of head-shaking, head-rolling, and head-turning in precopulatory situations.

In all observed cases, the male performed a single shaking movement immediately before he rushed toward the female and mounted her. Following treading the male performed a single rearing display, and then (in my observations) swam rapidly away while performing lateral head-turning movements. McKinney apparently observed some variations in postcopulatory behavior.

Displays of the Steller's eider (1–4) and long-tailed duck (5–8), including (1) male swimming normally past female, (2) female inciting, nearer male turning-of-the-back-of-the-head, (3) male display-preening, (4) upward-stretch, (5 and 6) male bill-toss and call (two views), and (7 and 8) male rear-end display (two views).

Nesting and brooding behavior. Judging from Brandt's comments (1943), male Steller's eiders normally remain in the vicinity of the nesting female for some time after incubation is underway and perhaps until the time of hatching. However, they apparently do not immediately begin their migration from the area; Percy (in Bannerman, 1958) observed coastal rafts of 400 to 1,000 birds, mostly male eiders, adjacent to the nesting grounds in late June. Since fewer than 5 percent of these were females, he believed that nesting success was apparently high in spite of dogs and other nest predators in the area.

Like the other eiders, females are very strong brooders and are extremely reluctant to leave their nests once incubation is underway. They tend to nest somewhat later than the other eiders and have a larger average clutch size, so that the period of hatching is likewise later than that of the other eiders. According to Blair (in Bannerman, 1958), when the ducklings are still quite young the females move their broods to the sea, where they often form "herds" and forage in the litter of tidal areas.

Postbreeding behavior. As Percy (in Bannerman, 1958) has pointed out, the unusually late molt of the flight feathers of this species allows it to undertake a fairly long migration to wintering areas prior to undergoing its wing molt and becoming flightless. Both sexes are represented among the flightless birds, although the sexes tend to segregate and occupy different parts of the bay.

Evidently in some instances this molt migration is in excess of 3,000 kilometers (Jones, 1965), but there seem to be yearly differences in the distances flown before molting. Thus, in some years the Steller's eiders arrive at Izembek Bay as early as August, while in others they have arrived as much as three months later, in early November. Peak numbers, however, usually do not occur until the eve of the spring migration. In 1977 the maximum number of adult males and subadults at Nelson Lagoon was more than 100,000 (Peterson, 1981), and nearly 80,000 were seen at Izembek Lagoon in 1980 (Laubham and Metzner, 1999).

Labrador Duck
Camptorhynchus labradorius (Gmelin) 1789

Other vernacular names. Pied duck

Range. Originally occurred along the northern Atlantic coast of New England and Canada's Maritime Provinces, probably mainly wintering along Long Island, but recorded from Labrador to Chesapeake Bay. Possibly bred in Labrador or farther north, but no definite breeding records were ever obtained. Todd (1963) reviewed the Labrador records and questioned the authenticity of some possible Labrador duck eggs, one of which is labeled "Labrador" (Glegg, 1951). Last reliably recorded in the fall of 1875.

Subspecies. None recognized.

Measurements. *Folded wing:* Males 210–220 mm, females 206–209 mm (Delacour, 1959).
Culmen: Males 43–45 mm, females 40–42 mm (Delacour, 1959).

Weights (mass). Audubon (1883) reported the weight of a male as 1 lb., 14.5 oz. (865 g) and of a female as 1 lb., 1 oz. (480 g). Chilton (1997) suggested that the seemingly low weight reported for the female might be an error, considering the similarity of the two sexes' wing and culmen measurements.

Identification

In the hand. *Adult males* had an entirely white head, neck, and scapular region except for a black stripe extending from the crown to the nape, a black collar around the base of the neck, and a yellowish area of stiffened feathers on the cheeks. The rectrices, upper tail-coverts, rump, back, primaries, and entire underparts were black, and the lower neck had a black collar. The upper and lower wing surfaces were white, except for the black primaries and their coverts; the scapular and tertials also were bordered with black. The postbreeding "eclipse" plumage of males is uncertain but might have been similar to that in the juvenile description (following). The iris color of adult males was probably reddish hazel (possibly yellow), and the legs and feet were probably grayish blue, but there is much disagreement on this too. The unusually soft-edged bill was black for most of its length, but the basal portion behind the nostrils was probably pale grayish blue and was separated from the black portion by a yellow, orange, or flesh-colored band.

Females were generally uniformly brownish gray, grading toward bluish slate dorsally, to sandy brown on the rump and tail coverts, and to light grayish brown on the underparts. The tail was very dark brown, the wing coverts were the same bluish slate as the mantle, except for the greater secondary coverts, which, with the secondaries, formed a white speculum. The primaries and their coverts were blackish brown, as in the male. The soft-part colors were probably similar to but duller than those of the male.

Juveniles apparently resembled adult females for most of their first year, with young males probably beginning to exhibit white feathers on the head, throat, and breast by the end of their first winter of life. One of the swimming birds in Peter Scott's Labrador duck watercolor might represent the first-year male plumage (he never explained its basis to me), but it corresponds fairly closely with Delacour's (1959) description of immature males and probably the male in eclipse plumage as resembling the female but grayer, but with some white feathers on the head and lower breast.

Natural History

Breeding distribution. The breeding range of this species is still uncertain. Chilton (1997) judged that it "probably nested along the Gulf of St. Lawrence and coastal Labrador; possibly on Ungava Peninsula and w. Canadian arctic." Of these, the western Canadian Arctic seems an unlikely breeding location, considering the currently available information and absence of reliable records from western Canada.

Winter distribution. Wintering occurred from Nova Scotia south to Chesapeake Bay, with most museum specimens having been obtained around Long Island, New York. Canadian winter records include Grand Manan Island, New Brunswick, and the Bay of Fundy (Chilton, 1997).

Habitat and foods. Evidently this little sea duck had a highly specialized diet, to judge from its unusual bill structure, which might have involved both dabbling at the surface in a shoveler-like manner and diving for its food. Since the birds were sometimes caught by fishermen on trotlines that had been baited with blue mussels (*Mytilus*), it might be imagined that mollusks were a part of their diet, and the birds apparently often fed close to shore along sandy bays or in estuaries where mussels might be abundant. It is quite possible that the Labrador duck occupied much the same habitat and consumed the same type of foods as the Steller's eider, which also has a soft-edged bill and is of very similar bodily proportions.

Social behavior. Nothing specific was ever recorded about the social behavior of Labrador ducks.

Reproductive biology. Nothing is known of their reproductive biology. Phillips (1922–26) suggested that the birds might have nested on a few islands in the Gulf of St. Lawrence, or in southern or eastern Labrador, in which case it would have been highly susceptible to nest robbing by fishermen or "eggers."

Status. Extinct since the mid-1870s, with the last known specimen taken in the fall of 1875, probably along Long Island. A less likely final record is from 1878, when a bird was reputedly shot near Elmira, New York, but the inland location makes this record suspect, and the specimen no longer exists. No convincing reasons for the bird's disappearance have ever been advanced; it was not an important sport species, nor was it sought after by market hunters. It seems most likely that a breeding-grounds disturbance, such as perhaps the arrival of effective mammalian predators such as foxes in a previously isolated nesting area, might have left the species defenseless, as was the case with the Aleutian cackling goose.

Labrador duck, male in breeding plumage

Epilogue

The Labrador duck has now been extinct for nearly 150 years, vanished forever along with the heath hen, Carolina parakeet, passenger pigeon, and an earlier America. It disappeared so swiftly and so quietly that it is not only difficult to compose a suitable epitaph but also impossible to write a complete obituary. We do not know for certain where it nested or exactly what it consumed, nor do we even have a record of the appearance of its downy young. Interred within the fewer than 60 skins, mounts, and bones that are scattered throughout the world's museums, like the deteriorating leaves of a now-dead oak, are the genes and chromosomes that represented the species' strategy for survival in a hostile world. That strategy failed, and in its failure the Labrador duck became the first of four waterfowl species doomed to extinction in historical times.

It may be fruitless to mourn for a bird that has been gone longer than the memory of any living human, but it would be folly to ignore this lesson of history. Uncounted millions of years of evolution failed to prepare the Labrador duck for survival in a world dominated by people with the ability to kill from great distances, to pollute the seas, and to ravage the wilderness. It is now well beyond the lifetime of any human since the Labrador ducks made their last ill-fated flights from their breeding grounds along the North Atlantic coast to the vicinity of Long Island; in that period our concern has gradually changed from the problems of how birds can survive modern humanity to the question of whether humankind can survive modern humans. The twenty-first century will no doubt provide that answer.

Harlequin Duck
Histrionicus histrionicus (Linnaeus) 1758

Other vernacular names. None in general use.

Range. Breeds in northern and eastern Asia, the islands of the Bering Sea, and in continental North America from Alaska and the Yukon south through the Cascade and Rocky Mountain ranges to Oregon and northwestern Wyoming. Also breeds disjunctively in northeastern North America, from Ungava Bay and northern Labrador south to central Quebec and southern Labrador, with isolated populations on the Gaspé Peninsula and Newfoundland. Also breeds on southwestern and southeastern Greenland. Winters in North America from the Aleutian Islands south along the Pacific coast to California, and on the Atlantic coast from southern Canada to New England.

Subspecies. None currently recognized.

Measurements. *Folded wing:* Delacour (1959): Males 200–210 mm, females 190–197 mm. Kuchel (1977): 12 males 180–201 mm, ave. 195.7 mm; 7 females 181–198 mm, ave. 187 mm.

Culmen: Delacour (1959): Males 25–28 mm, females 24–26 mm. Kuchel (1977): 12 males 27.5–29.9 mm, ave. 28.8 mm; 7 females 24.9–28.0 mm, ave. 26.3 mm.

Weights (mass). Nelson and Martin (1953): 5 males ave. 1.5 lb. (679 g), max. 1.6 lb. (725 g); 4 females ave. 1.2 lb. (543 g), max. 1.3 lb. (589 g). Kuchel (1977): 12 males ave. 683.2 g, range 660–710 g; 10 females ave. 626.5 g, range 545–660 g. Wright and Clarkson (1998): 141 males ave. 645 g, range 640–650 g; 61 females ave. 575 g, range 563–579 g.

Identification

In the hand. Recognizable as a diving duck on the basis of its large feet with lobed hind toes and lengthened outer toes. The combination of an extremely short, narrow bill (culmen length under 30 mm) and moderately long wings (190–210 mm) that are at least slightly glossed with purplish on the secondaries will eliminate all other species. Males in nuptial plumage are unmistakable; no other duck is predominantly slate blue with white spots and stripes. Females and dull-colored males have facial markings similar to those of female surf and white-winged scoters, both of which are substantially heavier (over 1,000 g) and have much longer bills (minimum 35 mm).

In the field. Normally found only along rocky coastal shorelines or on timbered and rapid mountain streams, harlequins are small diving ducks that appear quite dark on the water. Both males and females have white to grayish white areas on the cheeks, white between the eye and the forehead (continuous with the white

The North American breeding (hatched) and wintering (shaded) ranges of the harlequin duck. Regions of newly acquired or previously unrecognized breeding are shown by stippling.

cheeks in males, usually separate in females), and a rounded white spot halfway behind the eyes and the back of the head. Males might have additional white spotting, especially as they acquire their nuptial plumage, but these facial areas remain white to grayish white in all plumages. In flight, both sexes appear relatively dark, both above and below, exhibiting dusky brown under wing-coverts. When flying along mountain streams they remain quite low, following the course of the stream. When in coastal waters they forage in small flocks, often moving their heads in an elliptical fashion as they swim. Relatively silent birds, the male has a high-pitched, mouse-like squeal, and females utter a harsh croaking call.

Age and Sex Criteria

Sex determination. Sexing criteria based on feather characteristics have not been worked out, and apparently some first-year males are scarcely if at all separable externally from females. Older males have a more iridescent speculum, and the white feathers present on the head are bordered with black. When in full "eclipse" (basic plumage), males have considerably darker underparts than do females.

Age determination. Juveniles can be recognized for a time by their notched tail feathers as well as by their more spotted underparts and paler upperparts. After that, females cannot be obviously aged, but first-year males might be female-like, while second-year or older males apparently have the adult pattern. Examination of the oviduct or penis structure should serve to distinguish birds in their second fall of life from older, breeding birds. For details on ageing males, see Smith, Cooke, and Goudie (1998).

Distribution and Habitat

Breeding distribution and habitat. The North American breeding distribution of the harlequin duck is a curiously disruptive one, with the primary center in the forested mountains of western North America and a much smaller, more poorly defined secondary center in northeastern North America. In Alaska, harlequins probably nest throughout the Aleutian Islands (Murie, 1959). They are also common and are known to breed along the Alaska Peninsula, on Kodiak Island, on Kenai Peninsula, in the Copper River Valley, and along the coast of southeastern Alaska (Gabrielson and Lincoln, 1959). Although this essentially coastal region is no doubt their primary range, they do extend into the interior of Alaska and along the Bering Sea coast. There are relatively few interior or northern records of breeding, however. Harlequin broods have been seen regularly at Denali National Park, and eggs have been found at Loon Lake in the Brooks Range (Dennis Crouch, pers. comm.). The species is common during summer on St. Lawrence Island, and possibly breeds there (Fay, 1961).

In western Canada, harlequins reportedly nest in Yukon Territory, over much of British Columbia, and along the western edge of Alberta to the United States border. In British Columbia they are widespread, especially in coastal areas and the southern central interior, where they probably breeding from 250 to 2,500 meters of elevation (Davidson et al., 2008–2012). Harlequins also occur in summer in the Mackenzie River

Harlequin duck, male in breeding plumage

Valley, southeastward through Yukon Territory to perhaps Great Slave Lake, although breeding there is apparently unsubstantiated. In eastern Canada there are old breeding records from southeastern Baffin Island (Snyder, 1957) but no recent ones. Also breeds from the Ungava Bay coastline of Quebec and northern Labrador south to central Quebec and southern Labrador, and on the Gaspé Peninsula (Godfrey, 1986). Todd (1963) pointed out that, although harlequins are fairly common during summer on the coast of northern Labrador, no definite breeding records then existed. He did, however, provide a recent breeding record for the False River area of Ungava Bay. Harlequins also breed on the northeastern Gaspé Peninsula, northern New Brunswick, and in northwestern Newfoundland (Robertson and Goudie, 1999; Baldassarre, 2014).

South of Canada, harlequins are confined as breeding birds to the western mountains. They are probably commonest in Washington, breeding in the Olympic Mountains, on both sides of the Cascades, and in the Blue and the Selkirk mountains as well (Dennis Crouch, pers. comm.). In Oregon the harlequin has been found breeding in both the Wallowa and the Cascade mountain ranges (Gabrielson and Jewett, 1940). In California there are old records of breeding on a number of mountain streams, including the Merced, Cherry, Stanislaus, and Tuolumne Rivers, as well as at the headwaters of the San Joaquin River (Grinnell and Miller, 1944), but there are apparently no recent California records.

A breeding population also extends into the interior along the northern Rocky Mountains, along the Idaho–Montana border as far as Yellowstone National Park, where until at least the 1970s it was locally common below Fishing Bridge on the Yellowstone River, and it also occurs locally on mountain streams in Grand Teton National Park (Wallen, 1987). The harlequin's breeding range does not extend to Colorado, where it is a rare straggler (Bailey and Neidrach, 1967).

The preferred breeding habitat appears to be cold, rapidly flowing streams, often but not always surrounded by forests. There is apparently a limited attraction to tundra-like habitats, as indicated by the presence of breeding birds on Greenland and Iceland. Bengtson (1966) stated that no other European or North American duck is so closely bound to fast-running streams during the breeding season as the harlequin. He considered that the availability of suitable food, especially simuliids (blackflies), largely regulates the density and distribution of harlequins in Iceland, with nest site availability of secondary importance.

Wintering distribution and habitat. Large numbers of harlequins winter in the Aleutian Islands (Murie, 1959), and the birds are also common to abundant during winter along the bays of the Alaska Peninsula, the coastal waters of southeastern Alaska, Prince William Sound, Cook Inlet, and the Pribilof Islands (Gabrielson and Lincoln, 1959).

In Canada the harlequin winters along the coast of British Columbia, and also along the Atlantic coast in southern Labrador, Newfoundland, and the Maritime Provinces (Godfrey, 1986). Coastal British Columbia probably supports 12,000 to 15,000 wintering birds (Campbell et al., 1990).

In Washington, northern Puget Sound and the San Juan Islands probably harbor the largest number of wintering harlequins. In past years wintering flocks of 200 to 500 birds were reported (Jewett et al., 1953), but such numbers no longer are present. Along the open coastlines of Washington, Oregon, and northern California, harlequins might also be found during winter, particularly near rocky promontories. A few winter south to California, and stragglers extend to Baja California (Robertson and Goudie, 1999).

Along the Atlantic coastline, harlequins winter from the Canadian border southward along the rocky coastlines of Maine and beyond in diminishing numbers as these deep, rock-bound coasts give way to shallower, sandy, or mud-bottomed shores. More than half of the eastern population winters along coastal Maine (Robertson and Goudie, 1999).

The favorite wintering habitat of harlequins in Iceland has been described by Bengtson (1966) as those places where the surf breaks directly against the rocks, such as around the outermost peninsulas. They are usually found in waters 3 to 4 meters deep, and 100 to 300 meters from shore. They evidently prefer island

Harlequin duck, male in flight

points or other areas providing seclusion and protection from bad weather and the roughest water (Dennis Crouch, pers. comm.).

General Biology

Age at maturity. Plumage sequences indicate that the birds are mature in their second year of life (Bent, 1925). Hand-reared birds attained full plumage and began sexual display in their second winter (Charles Pilling, pers. comm.). Some females breed at two years of age (Reichel et al., 1997), and some males are able to obtain mates at two, but most do not form pair-bonds until they are at least three (Robertson and Goudie, 1999).

Pair-bond pattern. Bengtson (1966) reported that only 12 pairs were noted among a flock of 200 birds in late December, and that males desert their mates very shortly after incubation begins, indicating a yearly renewal of pair-bonds. However, there is some evidence of long-term pair-bonding in this species (Bengtson, 1972; Smith, Cooke, and Robertson, 2000).

Nest location. Bengtson (1966) reported that in Iceland harlequins prefer to nest on inaccessible islands, depositing their eggs in caves or holes in the lava, under dense bushes, or sometimes in rather open situations. No large trees presently occur in Iceland, and Bengtson partly attributed the species' use of other holes or crevices to the lack of hollow trees. Bengtson (1966, 1972) noted that the nests are always located very close to water, and the most prominent feature of the nest is that it is mostly protected from above by dense vegetation.

According to Dennis Crouch (pers. comm.), hole-nesting is not typical of North American harlequins; of 20 to 25 nest records he obtained, all were ground sites except for one involving an overturned stump in the middle of a stream. Robertson and Goudie (1999) also reviewed nest site preferences and stated the nest might be on the ground, on small cliff ledges, in tree cavities, or on stumps. There are records of nesting on the top of rotting stumps (Crowley, 1995), in low tree cavities (Cassirer et al., 1993), on the root mass of a fallen stump (Freeman and Goudie, 1998), in rock crevices (Bengtson, 1996; Crawley, 1994) and on cliff ledges and the tops of canyons (Bruner, 1997; Brodeur et al., 1998, 1999; Goudie, 1998; MacCallum and Bugera, 1998; Smith, 1999a). Nests are usually situated close to water, with ground nests often placed on low-lying midstream islands, or among woody debris. Thirteen of 16 Alberta nests were on islands (Smith 1998, 1999a), and 19 of 20 Oregon nests had vertical cover within two feet of the nest (Bruner, 1997). The width-to-length ratio of harlequin eggs averages about 1:1.42 and is less elongated that those of the surface-nesting long-tailed duck (averaging about 1:1.55) but not as rounded as the eggs of the cavity-nesting hooded merganser (1:1.24).

Clutch size. Bruner (1997) reported an average of 5.2 eggs for 21 nests in Oregon. Bengtson (1966) noted that 11 nests in Iceland ranged from 3 to 7 eggs, averaging 5.5. Records of 19 nests from across North America indicate a mean clutch of 6.2 eggs, a mode of 6, and a range of 4 to 8 (Dennis Crouch, per. comm.). Other samples from North America have means ranging from 5.2 to 6.1 eggs (Robertson and Goudie, 1999). According to Bengtson the egg-laying interval is surprisingly long: from 2 to 4 days, with 3 days between eggs probably normal.

Incubation period. Some early estimates of the incubation period were 31 to 34 days. Artificially incubated eggs have hatched in 29 to 33 days, but field observations indicate a period of only 27 to 29 days, averaging 28 (Bengtson, 1966, 1972).

Fledging period. Cassirer and Groves (1994) determined an average fledging period of 49 days for 12 Idaho broods, and Kuchel (1977) estimated a longer 55- to 56-day fledging span in Montana, but Wallen (1987) estimated an average of only 42 days for four Wyoming broods.

Nest and egg losses. Bengtson (1966) reported that in his study area the Arctic fox was absent and the mink was the only major mammalian predator. Egg-collecting activities by humans were then also a serious menace to harlequins but have since been prohibited. Avian predators of possible importance were parasitic jaegers, great black-backed gulls, and common ravens, but Bengtson did not estimate their relative importance as predators. The nests of this species are usually so well hidden that predators hunting visually would have difficulty locating nests, and mammals unable to reach nesting islands would also have limited effectiveness.

Bengtson (1972) estimated a high hatching success (87 percent) in Iceland, and other high rates of hatching success include 97.2 percent in Alaska (Crowley 1999) and 87 percent in Alberta (Smith, 1998). Nesting success was also high (74 percent) in an Alaskan study (Bruner, 1997). However, the incidence of nonbreeding among females is seemingly very high, and breeding success among those that do nest is low until the birds are at least 5 years old. Some varied estimates of nonbreeding incidences are 15 to 90 percent in Iceland, 14 to 26 percent in Alaska, 17 percent in Alberta (Robertson and Goudie, 1999), 47 to 50 percent in another Alaska study, and 53 to 66 percent in the Pacific Northwest.

The proportion of females successfully rearing one or more young is also highly variable from year to year, including yearly means of 7 to 55 percent in Montana, 15.9 percent in Alberta, 23 to 28 percent in Alberta, and 12.5 and 30 percent in two separate estimates in Alaska (Robertson and Goudie, 1999).

Juvenile mortality. Brood counts by Bengtson on early-age broods indicated an average brood size of 5.7 young, with an observed range of 4 to 10 ducklings. Duckling survival varied from 40 to 76 percent in different years (averaging 55 percent), and in four different study areas the mean number of young reared per female varied from 1.5 to 2.2, with food availability an important factor affecting reproductive success (Bengtson, 1972). Evidently minks and unfavorable weather were the primary mortality factors for young ducklings.

Duckling rearing success estimates (percent survival to fledging) in various North American studies include 13 percent in Alberta, 41 percent in Alaska, 55 percent in Oregon, 60 percent in Idaho, and 18 to 83 percent (three-year range of means) in Montana (Robertson and Goudie, 1999). Mean brood size at fledging among several studies included estimates of 2.4 young in Alaska, 3.3 in Oregon, 3.6 in Montana, and 3.9 in another Montana study (Robertson and Goudie, 1999). Assuming an initial brood size of 6, these figures would also suggest an average brood loss of about 40 percent prior to fledging.

Adult mortality. There are several estimates of annual mortality rates. One excellent example is that by Cooke et al. (2000). Based on 100 to 150 marked birds, the local survival rates were 82 percent for adult males, 56 percent for second-year males, and nil for juvenile males. In a Maine study, estimated annual survival rates were 86 percent for adult males, 66 percent for adult females, 52 percent for first-year males, and 48 percent for first-year females. In a study on Prince William Sound, Esler and Iverson (2010) calculated that annual survival of adult females averaged 86.6 percent over a three-year period, whereas in two subsequent years survival averaged 83.7 percent for adult females and 76.6 percent for first-year birds.

Harlequin duck, males and female foraging

General Ecology

Food and foraging. Cottam's (1939) summary of the foods found in 63 adults collected between January and September is the most complete analysis available. Virtually all of this food was of animal origin, with most of the volume consisting of crustaceans (57 percent), mollusks (25 percent), and insects (10 percent). Decapods such as smaller crabs, and soft-bodied crustaceans such as amphipods and isopods, appeared to be the favored types of foods, and together made up about half the total ingested volume. The mollusks included a surprising number of chitons, which are no doubt obtained from rocky shorelines in wintering areas, and a variety of gastropods probably found in similar habitats.

The only echinoderm of possible importance as a harlequin food is the spiny sea urchin (*Strongylocentrotus*), remnants of which occurred in nearly half the stomachs but in only very small quantities. Insect foods were more prevalent in summer samples and included species typical of rapidly flowing streams (stoneflies, water boatmen, and midge larvae).

In a tabular survey of six studies by Robertson and Goudie (1998)—based on wet mass, dry weight, or volume measurements—gastropod mollusks and decapod crustaceans were present in all six samples; bivalve mollusks were present in five; echinoderms, chitons, isopods, shrimp, and/or plants/algae were in four; polychaete annelids and/or fish eggs were in three; dipteran insects were in two; and barnacles, turbellarians (Annelida), sea anemones (Actiniidae), and sea cucumbers (Holothuroidea) were in one each.

As mentioned earlier, Bengtson (1966) considered the availability of midges (*Chironomida*, Diptera), blackflies (*Simulium*, Diptera), and caddisflies (Trichoptera) to be a determining factor in the abundance of harlequins on Icelandic streams and noted that the blackflies constituted the bulk of their nutrition on the breeding grounds. Similarly, birds collected during June in Siberia contained large numbers of caddisfly larvae as well as stonefly larvae and other insects, plus small fish remains (Dementiev and Gladkov, 1967).

In analyses of nine birds collected on coastal Maine in December, the amphipod *Gammerellus*, the gastropod *Nucella*, and the pelecypod *Lucinia* occurred in the largest numbers (Palmer, 1949). On their wintering grounds in coastal waters, harlequins often forage in heavy surf over shallow waters (Bengtson, 1966). Their dives there are thus usually of short duration and probably rarely exceed 20 to 30 seconds. On rivers they forage by skimming materials from the surface, diving, and up-ending, the last-named method apparently used only infrequently (Bengtson, 1966).

Bengtson reported that dives he timed on the River Laxa usually lasted 15 to 18 seconds, whereas Pool (1962) noted a range of 5 to 25 seconds on the same river. Bengtson calculated a dive-to-pause time ratio of 4:1 for the harlequin, 2.2:1 for the long-tailed duck, and 1.9:1 for the red-breasted merganser and Barrow's goldeneye foraging on the same stream. Thus he concluded that the harlequin duck is the most efficient diving duck species using rapid streams in Iceland. Pool similarly noted that harlequins dove with greater vigor and persistence than other species, and foraged in much stronger currents than the others.

Sociality, densities, territoriality. Bengtson (1972) noted that even on their nesting areas, harlequin territories were very indistinct, or sometimes seemingly lacking. He noted that possessive behavior of the male seemed to be related to his mate, rather than to a specific area. He also estimated that breeding densities averaged 1.3 pairs per kilometer of river (2.1 pairs per mile) and were highest near lake outlets.

This figure represents is a considerably higher breeding density than seems to be typical of the South American torrent duck (*Merganetta armata*), which is distantly related but occupies a similar ecological niche in torrential Andean streams (Johnsgard, 1966). Dennis Crouch (pers. comm.) observed harlequin densities in Washington state of one pair per 2 to 4 river miles, a figure much closer to the situation I found typical of torrent ducks.

Interspecific relationships. As Bengtson (1966) has pointed out, no other North American species of duck can effectively compete for food with the harlequin in its preferred habitats, fast-flowing streams. On wintering areas it consumes a variety of foods somewhat similar to those of the long-tailed duck, but the harlequin typically forages in areas with heavier surf and shallower waters than does the long-tailed duck.

Apparently there is little if any competition with other species of waterfowl for nest sites; Bengtson reported only that eggs of common mergansers have sometimes been found in the nests of harlequins. To what extent predators and parasites might play a role in the ecology of harlequin ducks is still rather uncertain. The harlequin's breeding populations are never so high nor are their nest sites so closely spaced or conspicuous as to attract predators in any numbers.

General activity patterns and movements. For the most part, harlequins appear to be daylight foragers, coming in each morning to favored ledges and rocky coves for foraging, and sometimes roosting on rocks in the evening (Cottam, 1939). During summer counts on the River Laxa, Bengtson (1966) found the highest incidence of foraging from 5:00 to 6:00 p.m., with a secondary peak in early morning. In July observations in Iceland, Pool (1962) reported that harlequins forage most heavily just prior to sunset. He also noted them to fly most actively then, as the birds flew up and down the river in small groups. Dennis Crouch (pers. comm.) observed foraging periods lasting from 6:00 to about 10:30 a.m., and from about 4:00 p.m. until dark.

Social and Sexual Behavior

Flocking behavior. Harlequins tend to fly in dense flocks, and they are highly sociable outside the breeding season (Bengtson, 1966). During winter they typically forage in groups of 5 to 25 birds, according to Bengtson. Other observers have estimated flock sizes as high as 500 individuals, but Dennis Crouch (pers. comm.) reported that groups of 5 to 6 are typical of western Washington.

Pair-forming behavior. Relatively few detailed observations of pair-forming behavior in harlequin ducks have been made, and the available descriptions have often not been in close agreement. Myres (1959a) observed only head-nodding as a social display and was not certain whether it was agonistic or sexual in function. He observed this movement in both sexes and sometimes heard a high-pitched note accompanying it.

Bengtson (1966) considered head-nodding to be the fundamental display movement and that it is basically an aggressive display, which is often followed by a threat posture. He also observed bill-dipping and associated lateral bill-shaking in males (often called water-twitching by others, and a frequent precopulatory display in other sea duck species), a wing-flapping that might represent a ritualized display, and both dorsal-preening and wing-preening movements that likewise were of uncertain display significance. Bengtson further noted that no "flight-display" is evidently present in this species, unless it occurs during winter and early spring when the birds are still at sea.

The only female display that he recognized in situations other than copulation was inciting. In this posture the female lowers her head and performs alternate head-turning movements, sometimes uttering a harsh call. Inciting has been seen much more rarely than one would predict if it plays an important role in pair formation. Gowans, Robertson, and Cooke (1997) and Inglis, Lazarus, and Torrance (1989) further added to these observations, noting that unpaired males will rush toward a female, in an apparent testing

Harlequin duck, male

of the pair-bond. Mate guarding of paired females by their mates is evidently an important mechanism for maintaining monogamous matings (Lazarus, Inglis, and Torrance, 2004).

Copulatory behavior. Bengtson's (1966) observations on 5 completed copulation sequences and 17 interrupted sequences provide the best description of copulation. He stated that the precopulatory behavior might be initiated by either sex, but usually the male, by commencing mutual head-nodding. In most of the sequences that Bengtson observed, mutual bill-dipping was also noted, and it seems probable that this element separates precopulatory head-nodding behavior from that seen in other situations. Finally, there is a precopulatory "rush" of the male toward the female, which might be repeated several times before mounting is achieved.

Evidently the female usually does not become prone until shortly before treading occurs. The postcopulatory behavior of both sexes is relatively simple and lacks specific posturing. Pearse (1945) provided an account of a single copulation sequence that likewise involved jerking movements of the head on the part of the male, but no other definite displays before or after copulation. Neil Smith (in Johnsgard, 1965) also observed attempted copulatory behavior in this species, which included rapid rushes toward the female.

Unpublished notes of Jay S. Gashwiler (pers. comm.) describe several observations of copulation or attempted copulation he observed during April and May in Oregon. One copulation sequence was preceded by a number of rushes toward the male by the female, and energetic "head-throwing" movements of the head back toward the shoulders. In a second instance, both sexes performed "head-throwing" and simulated pecking of the other's head or neck prior to copulation, and following treading the male chased the female over the water for a short distance a couple of times. Presumably these "head-throwing" movements are the same as the nodding movements described by others.

Nesting and brooding behavior. Bengtson's (1966) account of nesting behavior in this species is the most complete one available. He stated that females choose the nest site alone, although males closely follow them and stand "guard." The nest is simple, consisting of a thin layer of grass with occasional twigs and leaves, and lined with white down having reddish tips.

The female begins to incubate before the set is completed, and at that time begins to line the nest. She sits very tightly during incubation and probably leaves the nest for only very short periods at intervals of about 48 hours. Males leave their mates when incubation gets underway and begin to congregate in favored foraging areas. Following hatching, the female takes her brood to a secluded part of the river, moving about very little. The young of different broods sometimes merge, and in such cases are guarded by both females. Unsuccessful female breeders sometimes also participate in brood care. Apparently the young are not taken to the sea until they have fledged.

Postbreeding behavior. Bengtson (1966) stated that males remain at their areas of congregation for only a few days after deserting their mates and then depart for the sea. He mentioned, however, that observations of summer-plumage males on the River Laxa have been made. Thus, the degree of molt migration in this Icelandic population is uncertain. Some females apparently remain with their broods for a time after they fledge (Regehr et al., 2001).

In the North American population there are apparently certain areas favored by molting birds. Prince William Sound, the Strait of Hecate, and the Strait of Georgia all support large numbers of birds. Summer flocks, either of males or mixed sexes, have been reported around St. Matthew and St. Lawrence Islands and at Captains Bay, Unalaska Island (Gabrielson and Lincoln, 1959).

In eastern North America the northern tip of Labrador is a favored molting site, and several other locations in Labrador and Newfoundland are also used (Robertson and Goudie, 1999). Some birds from Labrador and Quebec molt in western Greenland (Boertmann, and Mosbech, 2002). Assemblages of drakes and immatures have also been reported from the Commander Islands, along the Russian coast, and at various other points (Dementiev and Gladkov, 1967).

Long-tailed Duck
Clangula hyemalis (Linnaeus) 1758

Other vernacular names. Oldsquaw, squaw duck

Range. Breeds in a circumpolar belt including all of Arctic North America, Greenland, Iceland, Arctic Europe and Asia, and the islands of the Bering Sea. Winters widely in the Northern Hemisphere on temperate and subarctic marine and fresh waters; in North America winters from Alaskan coasts south on the Pacific coast to Washington and infrequently beyond, on the Great Lakes, and on the Atlantic coast from Canadian waters south to South Carolina and rarely to Florida.

Subspecies. None recognized.

Measurements. *Folded wing:* Delacour (1959): Males 219–236 mm, females 202–210 mm. Palmer (1957): 12 males 225–237 mm, ave. 230 mm; 12 females 201–220 mm, ave. 211 mm.
Culmen: Delacour (1959): Males 26–29 mm, females 23–28 mm. Palmer (1957): 12 males 25–30, ave. 28 mm; 12 females 26–28 mm, ave. 27 mm.

Weights (mass). Nelson and Martin (1953): 31 males ave. 1.8 lb. (815 g), max. 2.3 lb. (1,042 g); 14 females ave. 1.4 lb. (634 g), max. 1.8 lb. (815 g). Schiøler (1926): 9 adult males (winter) ave. 750 g; 11 adult females ave. 686 g.

Identification

In the hand. Probably the most seasonally variable in appearance of all North American waterfowl (some feathers are molted three times per year), long-tailed ducks can be recognized as diving ducks by their lobed hind toe and long outer toe, and separated from other diving ducks by their short (culmen length 23–29 mm) flattened bill with a raised nail, almost uniformly brownish upper wing coloration, and white or grayish sides and underparts. White is always present around the eyes and can vary from a very narrow eye-ring to an extreme of an almost entirely white head.

In the field. Found only on deep lakes, large rivers, or along the coast, long-tailed ducks are fairly small diving ducks at home in the heaviest surf and the most bitterly cold weather conditions. On the water the birds appear to be an almost random mixing of white, brown, and blackish markings, but invariably the flanks and sides are white, or no darker than light gray, and some white is present on the head, either around the eyes, cheeks, or on the neck. Except during the summer molt, the elongated tail of males is also a good field mark, as are their black breasts. Individual females might be confused with female harlequin ducks, but they always have whitish rather than dark brown sides, and they might thus also be distinguished from immature

The North American breeding (hatched) and wintering (shaded) ranges of the long-tailed duck.

or female scoters. In flight, long-tailed ducks exhibit white underparts that contrast with their dark upper and lower wing surfaces. The courtship calls of male long-tailed ducks are famous for their carrying power and rhythmic quality, the commonest two sounding like *ugh, ugh, ah-oo-gah'* and *a-oo, a-oo, a-oo'-gah*.

Age and Sex Criteria

Sex determination. Sex and age criteria are still not clear, although adult males can be separated from adult females by their shiny black upper wing coverts (*vs.* blackish brown in females and immatures) and their rufous tertials and secondaries (*vs.* gray to rufous in females and immatures). Adult males also always have black breasts and (except during summer molt) greatly elongated tail feathers, plus pinkish color near the tip of the bill. Criteria for separating first-winter males from females include the male's presence of pink color on the bill, some blackish feathers on the breast, and some grayish white scapular feathers.

Age determination. Juvenile females are probably best recognized by the presence of notched tail feathers, whereas first-year males lack elongated tail feathers, have a mottled and imperfectly black breast, and have white scapulars that are not as long as in adults. Separation of second-year birds from adults might require examination of the reproductive tracts.

Distribution and Habitat

Breeding distribution and habitat. The long-tailed duck is probably the most Arctic-adapted of all ducks and has an associated breeding range in North America that extends from the northernmost parts of Ellesmere Island to the southern coastline of Hudson Bay. It also breeds extensively in Arctic latitudes of Eurasia.

In Alaska the long-tailed duck breeds from the base of the Alaska Peninsula in the vicinity of Ugashik, northward along the coastal tundra of the Bering Sea and Arctic Ocean, and into the interior along the valleys of the Nushagak, Kuskokwim, Yukon, and Kobuk Rivers, as well as at Denali National Park. It is probably the commonest breeding duck in such northern areas as Anaktuvuk Pass and the Colville Delta. Long-tailed ducks breed on the Aleutian Islands, and there are breeding records for St. Paul and St. Matthew Islands (Gabrielson and Lincoln, 1959). Breeding also commonly occurs on St. Lawrence Island (Fay, 1961).

In Canada the long-tailed duck is the most widely distributed duck throughout the Arctic regions (Snyder, 1957), commonly occurring along the coastlines of the Yukon, the Northwest Territories, Manitoba, Quebec, and Labrador. It also breeds on most and probably all of the islands in Canada's high-Arctic Franklin District, as well as on much of southwestern and most of northeastern Greenland. There may have once been an isolated breeding locality in northwestern British Columbia about 50 miles west of Atlin (Godfrey, 1986), but only three confirmed breeding records for the entire province exist, all at near-subalpine lakes (Davidson et al., 2008–2012).

The breeding habitat throughout this entire range consists of Arctic tundra in the vicinity of lakes or ponds, coastlines, or islands. Where shrubs are available for nesting cover, they are preferentially utilized, but grasses and sedges likewise are used; wooded country is apparently avoided.

Long-tailed duck, male in winter (breeding) plumage

Wintering distribution and habitat. As might be expected, the winter distribution of this species is about as widespread as its breeding distribution. It is common in the Aleutian Islands during winter (Murie, 1959), and it has been estimated that about 500,000 long-tailed ducks might annually winter around St. Lawrence Island (Fay, 1961). Along the southern coastline of mainland Alaska the birds are locally abundant, with great numbers occurring along the island channels of southeastern Alaska (Gabrielson and Lincoln, 1959).

In Canada, long-tailed ducks winter along the coastline of British Columbia, on the open waters of the Great Lakes, and along the Atlantic coast from southern Labrador and northern Newfoundland southward through the Maritime Provinces (Godfrey, 1986). They extend southward on the Pacific coast through Puget Sound and south along the open coastline of Washington, becoming uncommon in Oregon and relatively rare in California. Along the Atlantic coast they occur throughout New England and southward through the Chesapeake Bay region, where they are common residents, to about as far as the Carolinas, where they are uncommon (Pearson, 1919; Sprunt and Chamberlain, 1949).

In the Chesapeake Bay region they are fairly evenly distributed along the open ocean and coastal bays, the salt estuarine bays, and some of the brackish estuarine bays. A few are also seen on some of the fresh and slightly brackish estuarine bays (Stewart, 1962). The relatively great abundance of this species on Lake Michigan and some of the other Great Lakes indicates a stronger propensity for wintering on fresh water than is true of most eiders and scoters.

Because of their enormous Arctic breeding and wintering ranges, the total population size of long-tailed ducks is impossible to judge accurately. Estimates of up to 3 million to 4 million for North America have been made (Bellrose, 1980), whereas Goudie et al. (1994) estimated 2.5 million, and the Sea Duck Joint Venture (2007) recently estimated 1 million. Surveys of the breeding grounds since the 1990s suggest that a long-term decline in numbers has occurred since the 1990s (Baldassarre, 2014).

Long-tailed duck, male in summer (nonbreeding) plumage

General Biology

Age at maturity. Long-tailed ducks probably mature at two years of age, according to Ellarson (1956). Alison (1972) confirmed this with captive birds.

Pair-bond pattern. Not yet well studied, but the frequency of social display throughout the winter and spring suggests an annual renewal of pair-bonds. However, Alison (1972) reported that in a population of 95 pairs, at least 4 reestablished their pair-bonds in successive years.

Nest location. Bengtson (1970) reported on the locations of 348 long-tailed duck nests found in Iceland. He noted that this species tends to nest in pothole areas and also exhibits a distinct tendency to nest on islands when these are available. Of the total nests found, 202 were under low shrubs, 44 under high shrubs,

49 under sedge cover, 35 under the forb angelica (*Angelica* and *Archangelica*), and 18 were in mixed herb and grass meadows. Nests were usually quite close to water; the modal distance category from water was 3 to 10 meters.

Evans (1970) reported that long-tailed duck nests around Churchill, Manitoba, varied in their average distance from water depending upon the areas that were utilized for nesting sites. Those on islands in fresh water averaged about 2 meters, those placed on mainland beaches averaged 9 meters, and those on mainland tundra averaged nearly 30 meters from water.

Clutch size. Probably 6 to 7 eggs normally constitute the clutch. Jehl and Smith (1970) found an average clutch size of 6.3 for 17 completed first clutches at Churchill, Manitoba, while Alison (1972) reported an average of 6.8 eggs for 95 clutches. Bengtson (1971) reported that the average of 212 clutches from Iceland was 7.9 eggs, with significant yearly differences in mean clutch sizes that ranged from 7.0 to 8.4 eggs. He thought these differences were related to the relative abundance of chironomid (midge) larvae, with smaller clutches occurring during years when this food source was low. Twenty clutches by renesting birds averaged 6.0 eggs (Bengtson, 1972). The average egg-laying interval is 26 hours (Alison, 1972).

Incubation period. Incubation reportedly requires 24 days, or 24 to 25 days under bantam hens (Bauer and Glutz, 1969). Alison (1972) reported a 26-day incubation period under incubator conditions.

Fledging period. The fledging period is about 5 weeks (Bauer and Glutz, 1969), or 35 to 40 days (Alison, 1975). Alison (1972) noted that 19 captive-reared young fledged in 35 days, which represents one of the shortest of known waterfowl fledging periods, well adapted to high-latitude breeding. In comparison, the green-winged teal, the smallest of North American ducks (adults average less than 350 grams, or almost half the weight of long-tailed ducks), has a fledging period of about 42 days.

Nest and egg losses. Among a sample of 148 Icelandic nests observed during the egg-laying period, 48 percent of the nests were lost, with predation the most frequent cause. Among 55 normal-sized clutches that failed, desertion and predation by ravens and minks were the most frequent causes (Bengtson, 1972). Alison (1972) reported a 41 percent nest loss among 95 nests, with foxes and parasitic jaegers being the primary nest predators. A higher rate of nest losses (30.4 percent) occurred among island nests than mainland nests (20.5 percent), although nearly twice as many nests (59 versus 39) were placed on islands than on the mainland.

Juvenile and adult mortality. Frequent brood mergers prevent the use of brood size counts as a measure of prefledging losses. Boyd (1962) estimated an annual adult mortality rate of 38 percent (62 percent survival) for Icelandic long-tailed ducks. Other estimates of adult survival rates include 72 percent survival for Icelandic adults (C. S. Roselaar, in Cramp and Simmons, 1977) with a life expectancy estimate of 3.1 years, and a 75 percent annual survival for Alaskan birds (Robertson and Savard, 2002). Alison (1972) reported that the brooding-period mortality of adults at Churchill was nil for marked females, and only 1.5 percent for males over the 4-year study period 1968–71.

Hunting-related mortality is probably quite low for this species, as the birds are usually found well away from shore. However, Schamber et al. (2009) estimated a relatively low 74 percent annual survival rate for adult females on the Yukon–Kuskokwim Delta, based on 74 females followed from 1998 to 2004. This figure might have been influenced by a high level of local subsistence hunting.

General Ecology

Food and foraging. Cottam's (1939) study on long-tailed duck foods and foraging behavior is the most complete available for North America. He reported on the foods of 190 adults taken throughout most of the year, and of 36 juvenile birds collected during July. Both adult and juvenile birds had a predominance of crustaceans in the digestive tracts, with mollusks, insects, and fish in decreasing order of identified foods of adults.

Amphipod crustaceans (*Gammarus, Caprella*, etc.) alone constituted more than 15 percent of the adult foods, while phyllopod crustaceans (especially *Branchinecta*) totaled over 30 percent of the food volume of the juvenile birds examined. Adults also had consumed a substantial number of various crabs, shrimps, and other crustaceans, which together totaled nearly half of the food volume. Among the mollusks, bivalves, univalves, and chitons were all consumed but generally in rather small quantities. Insects were a fairly important source of food among birds collected during summer months, and most of the fish eaten were of little or no commercial value.

Ellarson (1956) reported that on Lake Michigan the long-tailed duck and whitefish populations are both closely dependent on amphipods (*Pontoporeia*), thus accounting for the high gill net mortality found there. Lagler and Wienert (1948) had earlier reported on the predominance of these amphipods and a small bivalve (*Pisidium*) in a sample of 36 birds from Lake Michigan. Apparently the long-tailed ducks wintering in Danish waters have a higher dependency on mollusks; Madsen (1954) reported that the volumetric analysis of 110 birds revealed 65 percent were bivalve mollusks, 8 percent univalve mollusks, 27 percent crustaceans, along with a small amount of other animal foods.

In a 12-study comparison of foods found in the digestive tracts of molting and wintering long-tailed ducks, bivalves and gastropods were reported in eight studies; amphipods in seven; fish and fish eggs in five; shrimp, isopods, and decapods in four; mysids (opossum shrimp), oligochaete annelids; and insects in two; and daphnids, anostracods, and cumaceans (hooded or comma shrimp) in one (Robertson and Savard, 2002).

The long-tailed duck is famous for the depths it sometimes reaches during foraging, with many reports of birds foraging at 50 to 100 feet and a few records of individuals apparently exceeding 150 feet in their dives (Cottam, 1939; Schorger, 1947, 1951). Probably the normal foraging depth is no more than 25 feet, at least in coastal areas where the birds are foraging on mollusks and other invertebrates of the subtidal zone. Lagler and Weinert noted that most of the birds caught in gill nets on Lake Michigan are taken at depths

Long-tailed duck, male (summer)

of 8 to 16 fathoms, and that the greatest abundance of their two primary food species (*Pisidium* and *Ponto poreia*) occurs at depths of less than 60 meters (33 fathoms).

Sociality, densities, territoriality. Long-tailed ducks are not considered social nesters, and Bengtson (1970) reported that this species, like most other waterfowl he studied, showed an essentially random distribution of its nest sites. He observed a tendency to select islands for nest sites, but the overall average density of long-tailed duck nests per square kilometer on 13 study areas was 44 nests (6.9 acres per nest).

In contrast, Alison (1972) found a breeding density of 4.5 pairs per square mile during four different years; he also noted positive male territoriality, with males defending a small pond, or part of a larger pond. Although the males held territories of varying sizes, and sometimes the same territory in different years, the females were nonterritorial and rarely nested in their mate's territory. Instead, they nested in a semicolonial manner, with nearly two-thirds of the nests within 100 feet of at least one other active nest.

Long-tailed duck, male (winter)

Interspecific relationships. In their seeming concentration on soft-bodied crustaceans, such as amphipods, and their secondary utilization of mollusks, long-tailed ducks probably only actively compete for food with harlequin ducks and, perhaps locally, Steller's eiders. Considering the differences in the geographic distributions and preferred habitats of these species, it seems likely that there is little actual competition among them. Phillips (1925) noted that long-tailed ducks rarely associate with other ducks, and Mackay (1892) mentions seeing them a few times in the company of eiders. Hull (1914) observed that in Jackson Park, Chicago, long-tailed ducks avoided and were avoided by the other common wintering diving ducks, such as scaup and goldeneyes.

Long-tailed ducks build well-concealed nests that are notoriously difficult to locate. Even so, mammalian predators such as foxes are responsible for some nest losses, as are jaegers, and larger Arctic gulls no doubt account for the loss of some young as well.

Evans (1970) studied the nesting association of long-tailed ducks and Arctic terns at Churchill, Manitoba, and summarized evidence that ravens and parasitic jaegers sometimes are important egg predators, and the latter might also consume ducklings. Evans reported that the long-tailed ducks he studied apparently gained protection from nesting near Arctic terns, and he suggested means by which positive nesting associations between the two species might gradually develop. However, Alison (1972) questioned whether this relationship is actually beneficial to the long-tailed ducks, since he did not find any lowered nest predation rates for nests in tern colonies but did confirm Evans's observations that long-tailed ducks often nest in association with Arctic tern colonies.

General activity patterns and movements. As with the other marine ducks, the general pattern of activity is one of foraging during the daylight hours in fairly shallow waters and moving to deeper bays or the open ocean for nocturnal resting. Mackay (1892) described this pattern in New England. He noted that in the Nantucket area the birds forage during the day in waters some 3 to 4 fathoms deep and start to leave about 3:00 p.m. for deeper waters. The afternoon flight continues until after dark. Apparently the birds sometimes remain on their feeding grounds after dark on clear, calm nights, but most birds shot during early morning hours have empty stomachs. In rare cases where the ducks have been seen foraging on freshwater ponds near the coast, they fly in to these ponds early in the morning and return to the coast about sunset. Alison (1970) observed a similar nighttime movement to deep water on Lake Ontario.

Social and Sexual Behavior

Flocking behavior. Although long-tailed ducks rarely associate with other species, they do form large single-species flocks, especially during fall. Mackay (1892) mentioned that flocks arriving in New England during fall usually were in groups of 75 to 100 birds, but sometimes flocks of more than 1,000 could be seen on these wintering areas. This would seem to represent an unusually large flock, however. Dementiev and Gladkov (1952) reported that fall flocks of up to 1,500 birds have been reported, that most wintering birds are found in small bands of up to 15 birds, and that spring migrant groups might number 300 to 400 individuals. The fact that these birds often feed in unusually deep waters, well out from the shore, makes it relatively difficult to obtain counts of flocks.

Pair-forming behavior. Pair-forming behavior begins on the wintering areas, sometimes as early as December (Alison, 1970). By early May, about 70 percent of the adult males Alison observed near Toronto were already paired. The loud calls of the males, associated with social display, make courtship activity highly conspicuous.

Male displays of the long-tailed duck (1) and common scoter (2–9), including (1) wing-flapping, (2) neck-stretching and calling, (3) breast-preening, (4) forward-stretch, (5) head-shake, (6) upward-stretch, and (7–9) tail-snap–low-rush sequence.

Myres (1959a) was the first to provide a partial description of long-tailed duck sexual behavior patterns, and my observations (1965) and those of Alison (1970) have supplemented his observations. Myres recognized two displays associated with these calls, the "bill-toss" and the "rear-end" displays. During the former call, which sounds like *ugh, ugh, ah-oo-gah'*, the head might be quickly tossed backward beyond the vertical while the hindquarters are maintained in a normal position. In the rear-end display, the neck and head are extended downward and forward toward the water, while the tail is raised to a nearly vertical position, and a loud call, *a-oo, a-oo, a-oo'-gah*, is uttered.

Although not noted by Myres, Alison (1970) and I have observed neck-stretching and a turning-of-the-back-of-the-head by males as apparent displays, and a wing-flapping that possibly also represents a form of display behavior. Alison has observed a number of additional male displays, including lateral head-shaking, "porpoising," "steaming," "breast display," a short flight or "parachute display," and others. The most common female display is a chin-lifting that is probably a type of inciting, but Alison has observed some additional female postures as well.

Copulatory behavior. Myres (1959a) observed three instances of copulatory behavior but noted no specific associated displays. Alison (1970) reported that in six precopulatory situations the males invariably performed bill-tossing and lateral head-shaking, while bill-dipping, neck-stretching, and porpoising were also observed in some cases. In all cases the females performed lateral head-shaking and neck-stretching; a prone or soliciting posture was also sometimes observed. Various male postcopulatory displays were observed, including bill-tossing, neck-stretching, head-shaking, turning-of-the-back-of-the-head, and a sequence of porpoising, head-shaking, and wing-flapping.

Nesting and brooding behavior. According to Alison, the female hollows out her nest site immediately prior to the laying of the first egg. She normally sits very tightly (I have often been able to approach sitting females within arm's reach) but typically feeds twice a day, and on warm days might leave the nest for several hours. The males abandon their mates when the hens begin incubation and either remain in the general area to undergo their molt, or completely leave the area.

After hatching, the ducklings are often reared on freshwater ponds or lakes, but sometimes they are taken to salt water when they are only a few days old (Phillips, 1925). The female typically leads the brood to open water rather than to shore when the young are threatened, and as they grow older they gradually move from smaller sedge-lined lakelets to larger reservoirs and marine waters (Dementiev and Gladkov, 1967).

Postbreeding behavior. Throughout most of its breeding range, the breeding long-tailed ducks undergo their postnuptial molt in the breeding area, either as solitary birds or in small flocks. However, large concentrations of molting birds occur around St. Lawrence Island, and on the western and northern coasts of Alaska. Notable molting sites (totaling 30,000–40,000 birds) occur between Prudhoe Bay and Demarcation Point, such as around Herschel Island (5,000–10,000 birds), and at Simpson Lagoon (west of Prudhoe Bay), where 51,000 birds have been reported (Baldassarre, 2014).

Long-tailed duck, female incubating (Manitoba)

Other known molting sites include the north shore of mainland Nunavut, southeastern Southampton island, Ungava Bay, and Hudson Strait (Palmer, 1976). In eastern Siberia the breeding males and immatures evidently undertake an extensive molt migration to Wrangell Island prior to molting (Salomonsen, 1968).

Fay (1961) mentioned seeing considerable numbers of flightless long-tailed ducks on the lagoons and lakes of St. Lawrence Island, in flocks of fewer than 10 to more than 100. Whether these are immature nonbreeders that never left their wintering grounds or are birds that have moved in from other areas is apparently unknown. Probably the latter is the case, since Alison (1970) observed that the migration of the sizable wintering population in the Toronto area is always total.

Black Scoter
Melanitta americana (Swainson) 1831

Other vernacular names. American scoter, common scoter, coot

Range. Breeds in North America from western and northern Alaska east to northwestern Canada (Yukon and Northwest Territories), mostly in boreal forest and parklands, also in northern Ontario, Quebec, and Labrador. Winters in North America along the Pacific coast from the Pribilof and Aleutian Islands south to southern California, on the Atlantic coast from Newfoundland south to about South Carolina, and to a limited extent in the interior, especially on the Great Lakes. In 2010 the American Ornithologists' Union specifically separated *Melanitta americana* from *M. nigra* (the Eurasian common scoter), which breeds in Iceland, the British Isles, northern Europe, northern Asia, and islands of the Bering Sea, and winters widely in Europe and Asia.

North American species. *M. americana* (Swainson): Black Scoter. Breeds and winters in North America, as indicated above.

M. nigra (L.): Common Scoter. Accidental in Greenland and continental North America during winter. Breeds from Iceland eastward throughout northern Europe and northern Asia.

Measurements (*M. americana*). *Folded wing:* Palmer (1975): Males, 229–241 mm, ave. of 12, 234 mm; 11 females, 206–230 mm, ave. 222 mm.

Culmen: Palmer (1975): Males, 40–47 mm, ave. of 12, 43.3 mm; females, 39–44 mm, ave. of 11, 42.2 mm. Bordage and Savard (1995): Males ave. 43.7 mm, females ave. 40.9 mm.

Weights (mass). Bordage and Savard (1995): 34 *M. americana* adult males ave. 1,117 g; 21 females ave. 987 g. Nelson and Martin (1953): 8 *M. americana* males ave. 2.4 lb. (1,087 g); 4 females ave. 1.8 lb. (815 g); max. 2.8 lb. (1,268 g) and 2.4 lb. (1,087 g), respectively. Schiøler (1926): 5 adult *M. nigra* males ave. 1,164 g; 7 second-year males ave. 1,101 g; 11 juveniles ave. 1,084 g; 6 adult *M. nigra* females ave. 1,055 g; 7 second-year females ave. 1,070 g; 4 juveniles ave. 876 g.

Identification

In the hand. Recognizable in the hand as a diving duck by its enlarged hind toe and lengthened outer toe. The unusually narrow outermost primary (less evident in juveniles) and the relatively long (80–100 mm) and rather pointed tail (with 16 rectrices) identify both sexes as black scoters. The bill is not feathered on the lateral surface or dorsal culmen, and no white feathers are anywhere on the body except in juveniles, which have whitish underparts. The black scoter is separable from the similar-sized surf scoter by the former's lack

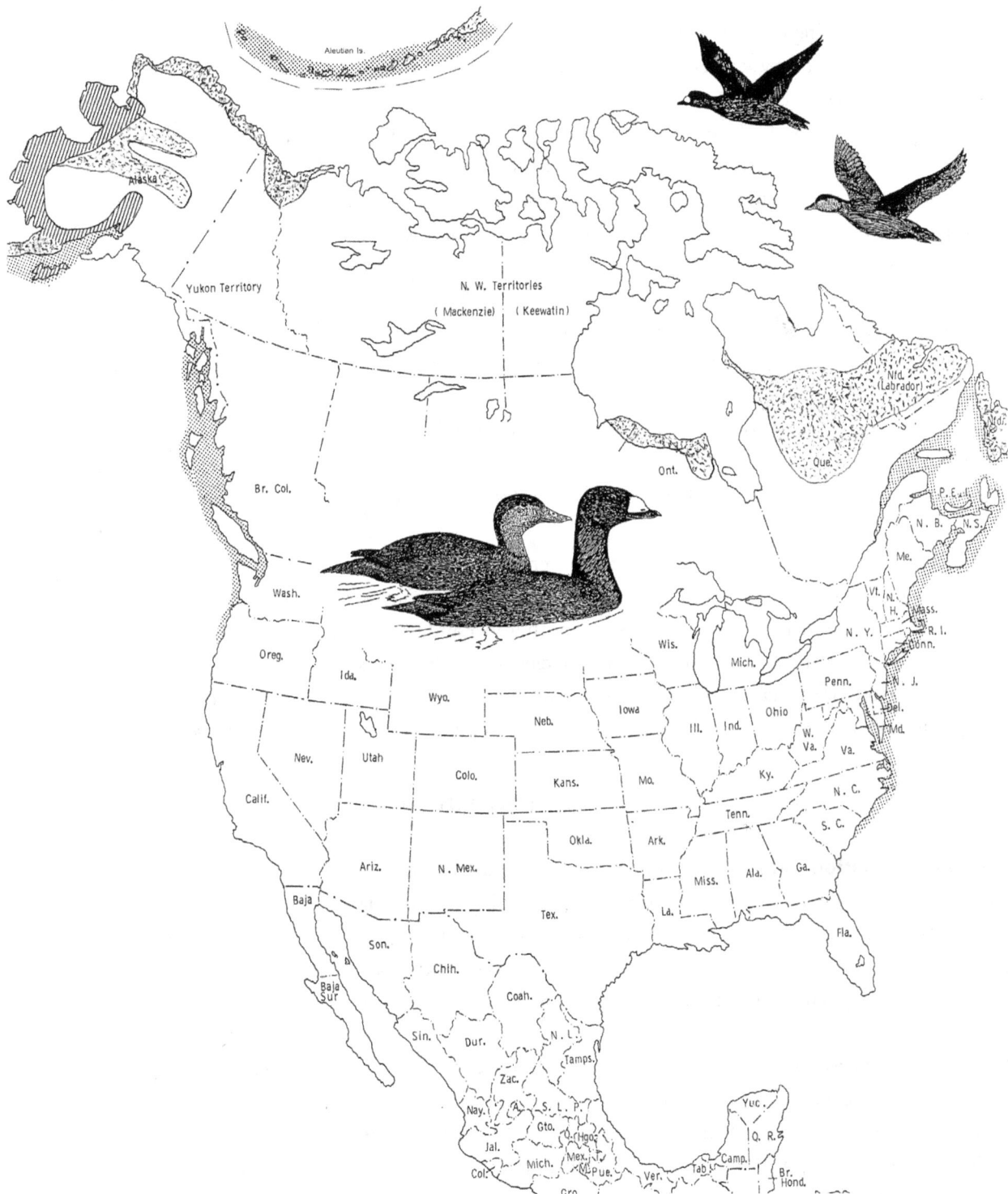

The North American breeding (hatched) and wintering (shaded) ranges of the black scoter. Regions of newly acquired or previously unrecognized breeding are shown by stippling.

of feathers extending down the forehead almost to the nostrils, and from the white-winged scoter by absence of white secondary feathers. Both of these latter species usually have 14 rectrices.

Adult black scoter (*M. americana*) males have bill knobs that are notably larger than in European common scoters (*M. nigra*) and are entirely yellow rather than black laterally. Wing and culmen measurements of *M. nigra* and *M. americana* overlap completely and are not useful in distinguishing them, but the tip of the bill is more hooked in *americana* than in *nigra*, and the male courtship call, a mellow whistle in both species, is substantially longer in duration in *americana* (Sanger, 1999). Dean and the British Birds Rarities Committee (1989) provided additional information on distinguishing these very similar species.

In the field. Black scoter males are the blackest of all North American ducks, and females are the most uniformly dark brown of all these species. The best field mark for mature males, other than their overall black color, is a yellow enlargement at the base of the bill, which is much larger in the black scoter than in the common scoter. Females of both species have a two-toned head and neck pattern: dark brown above and grayish white on the cheeks, throat, and foreneck. Juveniles are similar, with an even sharper contrast in their head pattern. Black scoter females and juveniles are probably not separable from those of the European common scoter in the field.

All scoters take flight by running over the water, and they fly rather low but swiftly near the surface. The wings of male scoters of both species also produce a strong whistling noise in flight. Both black and common scoters appear dark brown or blackish on both upper and lower surfaces and have no white on the head or wings. The call of courting males of both species is a mellow whistle, but the courtship call of the male American scoter is substantially longer in duration than is that of the common scoter (Sangster, 2009). Females of both species utter grating notes reminiscent of a door swinging on rusty hinges.

Age and Sex Criteria

Sex determination. Adult males can be readily separated from females and immatures by their entirely black plumage. The tenth primary (outermost, excluding the vestigial eleventh primary) has its inner vane strongly narrowed for about 6 centimeters, or nearly the entire exposed length, while the corresponding feather of females is less strongly narrowed for only about the distal 4 centimeters. In first-year males this feather gradually tapers in width toward the tip or is slightly narrowed toward the tip.

Age determination. Probably first-year females can be distinguished from older ones by the gradually tapered inner vane on their tenth primary and by their more whitish cheeks; in older females the distal half of the inner vane is only about half as wide as the proximal half, and the cheeks and throat are a darker shade. First-year males are quite female-like, and their tenth primary gradually tapers toward the tip. Some black feathers are acquired on the upperparts, but the abdomen and wings remain brownish (Dwight, 1914). By their second year, males are apparently indistinguishable from older birds.

Black scoter, male in breeding plumage

Distribution and Habitat

Breeding distribution and habitat. The North American breeding distribution of the black scoter is still the least understood of all the North American sea ducks. It is centered in Alaska, along the Bering Sea coast, and it also breeds in Russia, east to at least the Lena River Delta, where it is reportedly in contact with the common scoter. The black scoter evidently breeds on the Aleutian Islands, although in still uncertain numbers, and is an abundant breeder on the Alaskan Peninsula, with perhaps 15,000 breeding birds (Stehn et al., 2006).

Stehn et al. (2006) estimated a total Alaskan breeding population of more than 100,000 birds, with major concentrations on the Yukon–Kuskokwim Delta (53,000), Bristol Bay (25,000), and fewer numbers on the Arctic Coastal Plain, Selawik, and the Seward Peninsula. It has been reported breeding in the Bristol Bay region along the Kvichak River, on Nelson Island, and along the mouth of the Yukon River, where it sometimes has constituted up to 25 percent of the observed waterfowl. Farther north, it nests at Cape Prince of Wales and the Shishmaref region of the Seward Peninsula. In the interior of Alaska the black scoter has been reported nesting in Denali National Park and in the vicinity of Lake Louise, which is located on a tributary of the Sustna River (Gabrielson and Lincoln, 1959).

In Canada, earlier records of definite breeding were very limited in the 1960s. They included the Anderson River Delta, the Windy River area of southern Keewatin District, Leaf Bay in northern Quebec, and various localities in Newfoundland (Godfrey, 1986). It is now known that the main population centers are along the Hudson Bay lowlands of Ontario, Quebec, Labrador, and Newfoundland. Population centers include the Great Whale River and Lake Bienville in northern Quebec, interior wetlands along the eastern shoreline of Hudson Bay, central Labrador, and south-central Newfoundland (Bordage and Savard, 1995). Recent Canadian surveys (Canadian Wildlife Service Waterfowl Committee, 2013) have shown that breeding extends well to the west of Quebec, in northwestern Ontario, northern Manitoba, Nunavut, and east of Great Bear Lake in the Northwest Territories. Atlantic Flyway spring counts during 2005 revealed about 90,000 black scoters, with 52,000 in the Chaleur Bay and nearly 49,000 in the St Lawrence Estuary.

Black (and common) scoter breeding habitat seemingly consists of freshwater ponds and lakes in tundra or slightly wooded country. Black scoters prefer shallow, rocky shoreline lakes of medium size in tundra or in the boreal forest–tundra transition zone. In Iceland the common scoter prefers to nest in pothole areas where shrubs are present (Bengtson, 1970). Since shrubs are an evidently favored type of scoter nest cover, it would seem that shrubless lowland tundra does not represent ideal habitat. Further, the extremely late nesting of scoters would tend to prevent them from extending far into high Arctic tundra habitats.

Wintering distribution and habitat. Black scoters winter commonly along the Aleutian Islands and the Alaska Peninsula from Kodiak Island to Atka Island (Murie, 1959). They have been reported as abundant at Attu Island, but Kenyon (1961) did not list them for Amchitka. They also winter among the islands and channels of southeastern Alaska (Gabrielson and Lincoln, 1959) and along coastal British Columbia southward to Puget Sound, where they are the least common of the scoters (Yocom, 1951). Additionally, they extend in winter along the coast southward into Oregon and occasionally to southern California.

On the Atlantic coast black scoters winter from Newfoundland southward, with a few occurring irregularly on the Great Lakes (Godfrey, 1986). They are frequently quite common as far south as Chesapeake Bay, where they are generally found in the littoral zone of the ocean, with a few occurring on coastal bays and occasionally on salt and brackish estuaries of the bay itself (Stewart, 1962). Farther south they are generally the least common of the three scoter species and perhaps normally range as far south in winter as South Carolina (Sprunt and Chamberlain, 1949), and rarely to the Gulf of Mexico.

As Stewart has mentioned, the optimum winter habitat for this and the other scoters is the littoral zone of the ocean, usually within a mile of shore and in the area just beyond the breakers. There they both forage and rest, relatively independent of tidal action and human disturbance. Generally the black scoter seems to prefer areas where the water depth does not exceed 25 feet and where mussels can be found in large quantities (Cottam, 1939).

General Biology

Age at maturity. No definite information exists regarding age at maturity except that the fully adult plumage of the male is not attained until its second fall of life, and perhaps the bill coloration and enlargement might not be fully developed until even later (Dwight, 1914). Thus, pending evidence to the contrary, breeding at the end of the second year of life would seem most probable (Palmer, 1976).

Black scoter, pair in flight

Pair-bond pattern. Pair-bonds are evidently renewed yearly. Bengtson (1966) noted that many female common scoters were already apparently paired on their arrival at Icelandic breeding grounds, but courtship activities were frequent during May and June, and some active, unpaired females were seen as late as mid-June. Evidently the males deserted their mates and quickly left the area as soon as the females began to incubate.

Nest location. In Iceland, common scoter nests are usually placed under a dense cover of birch- and willow-scrub (Bengtson, 1966). Of 308 nests found by Bengtson (1970), 199 were under low shrubs, 78 under high shrubs, 12 were in holes, 11 under angelica forbs (*Angelica* and *Archangelica*), 5 were in meadows, and 3 among sedges. Nests were frequently situated in locations 10 to 30 meters from the nearest water. There was no evident tendency toward nesting on islands, and indeed the relative nest density was somewhat lower on islands than on the mainland.

Clutch size. So few clutches of black scoters have been found that it is difficult to know what is a typical clutch size for this population. Twenty Alaskan nests averaged 7.7 eggs (C. Dau, cited in Bordage and Savard, 1995). The European common scoter similarly produces clutch sizes of 6 to 9 eggs, with occasional records of 5 and 10, with one (very probably multiple) clutch of 13 eggs reported (Bauer and Glutz, 1969). A sample of 187 initial Icelandic clutches averaged 8.74 eggs, with significant yearly differences in mean clutch sizes that ranged from 7.56 to 9.04 (Bengtson, 1971). Thirty renesting efforts averaged 6.1 eggs (Bengtson, 1972).

Incubation period. The incubation period was reported by Delacour (1959) and Johnstone (1970) for European common scoters as 27 to 28 days. Also estimated as 30 to 31 days for Russian common scoters by Dementiev and Gladkov (1967), and as 27 to 31 days by Cramp and Simmons (1977), these data collectively average 29 days. Comparable data for North American birds are still lacking.

Fledging period. The fledging period is estimated by Lack (1968) for European common scoters as 6.5 weeks. Similar estimates have been made for the European population by others, such as 45 to 50 days by Cramp and Simmons (1977), but no estimates are yet available for North American birds.

Brood rearing habitats of black scoters consist of small (under 10-hectare) and medium-sized (10- to 100-hectare) lakes, mostly without shoreline vegetation. The same brood habitats are used by surf scoters, but black scoters additionally use small ponds and lakes having riparian grass and sedge vegetation (Décarie et al., 1995). Brood amalgamation is common during the later stages of brood growth.

Nest and egg losses. Seventeen percent of 109 common scoter nests were lost during the egg-laying period, primarily through predation. Of 19 clutches that failed to hatch, predation (11 clutches) and desertion (6 clutches) were major factors, and 2 clutches were flooded. Most of the eggs that failed to hatch (86 percent) were infertile (Bengtson, 1972).

Juvenile and adult mortality. Both prefledging and postfledging mortality losses of juvenile black and common scoters are little studied. Over four years of study, common scoter duckling survival in Iceland varied from 20 to 60 percent, with most duckling losses occurring during the first two weeks after hatching, and later-hatched broods survived less well than earlier ones (Bengtson, 1972).

Boyd (1962) estimated the annual adult mortality of Iceland's common scoter population as 33 percent (67 percent survival), and other Icelandic data suggest a 77 percent survival (Bordage and Savard, 1995). A less reliable estimate for Russian birds was 80 percent annual survival, and a maximum known longevity of eight years for wild birds (Bianki, 1992). No estimates are yet available for annual survival rates in black scoters.

Common scoter, male and female

General Ecology

Food and foraging. Cottam's (1939) study of the foods taken by 124 adult black scoters collected during ten months of the year is the most comprehensive available for North America, while Madsen's (1954) study of 219 samples from Danish coastal waters provides comparable information for the common scoter. Cottam reported that nearly 90 percent of the volume of food present was of animal origin, with mollusks constituting most of the animal foods. The most important of these were blue mussels (*Mytilus edulis*) and related forms, with short razor clams (*Siliqua*) of secondary importance. A larger consumption of barnacles was indicated than appears typical of other scoter species.

Madsen similarly found that mollusks constituted 77 percent of common scoter food by volume, of which bivalves (especially *Mytilus edulis*) account for the majority, while polychaete worms, crustaceans, and echinoderms made up the remainder. The four most important foods were apparently blue mussels, cockles (*Cardium*), univalve mollusks (*Nassa*), and tubeworms (*Pectinaria*). Thus, although birds in both species' populations primarily consumed mussels and other mollusks, the consumption of crustaceans, annelids, and other invertebrates seemingly varied with availability or other factors.

In a three-study comparison of foods found in the gizzards of black scoters, bivalve mollusks comprised by far the greatest volume, followed by crustaceans (barnacles), and lastly polychaete annelids. Blue mussels and other bivalves of the mussel family (Mytilidae) were greatest in volume, followed by razor clams (*Silique*) and various other clams (*Mesoderma, Tapes, Protothaca, Mya*) and oysters (*Ostrea*) (Bordage and Savard, 1995).

Plant materials play a small role in the diet of scoters, and even among summer samples Cottam reported that only about 13 percent of the identified food materials were of this source. Cottam reported that black scoters usually forage in water less than 25 feet deep, but they have been known to forage in waters 40 feet in depth. He also noted that the European species has been reported to forage primarily in waters 6 to 12 feet deep. Salomonsen (1968) mentioned that molting common scoters in the North Sea mostly occur in waters less than 5 meters deep.

Sociality, densities, territoriality. Judging from the data of Bengtson (1970), the nests of common scoters are essentially distributed at random, with no tendencies toward aggregation in nesting colonies. The average nesting density that he reported on 13 different study areas in Iceland was 53 nests per square kilometer (5 acres per nest). There seems to be no evidence relating to the possible existence of definable territoriality in this species, although males defend the area immediately around their males. Aerial surveys over large areas in Quebec and Labrador suggest maximum breeding densities of 12 pairs per 100 square kilometers (Bordage and Savard, 1995).

Little is known of overall population sizes in North America. Bordage and Savard (1995) estimated a maximum of 25,000 broods in Alaska, and a breeding population of about 50,000 pairs. Savard and Lamothe (1991) estimated that the Quebec and Labrador populations might total 43,500 pairs. Other more optimistic but older estimates have included 235,000 birds in Alaska and 315,000 elsewhere in North America (Bellrose, 1980). Goudie et al. (1994) reported a ten-year (1984–94) declining trend of scoter populations in western North America, and scoter surveys back to the 1950s confirm this long-term scoter decline (Canadian Wildlife Service Waterfowl Committee, 2013). Large-scale scoter population surveys cannot separate black scoter numbers from those of other species, but it is certainly the rarest of the three scoters.

Interspecific relationships. To what extent competition for food might exist between the black scoter and the other two species of this genus is unknown. All three forage predominantly on mollusks, but the size differences of adults (surf scoter smallest, white-winged largest) might bring about differences in effective foraging depths. The black and surf scoter seemingly both rely heavily on blue mussels and related species, but the surf scoter generally forages closer to the coastline than does the black scoter (Cottam, 1939). Phillips (1926) noted that, although the black scoter is most often seen in single-species flocks, it more commonly associates with surf scoters than with white-winged scoters.

General activity patterns and movements. Phillips reported that this species is relatively active throughout the day, and that migratory or local movements can occur at almost any time of day. According to him, the birds normally move into shallower waters for foraging early in the morning, often coming from some distance.

Social and Sexual Behavior

Flocking behavior. Apart from the obvious fact that migrant and wintering flocks of black scoters are often extremely large, there seems to be little specific information on flock sizes of this species. Atkinson-Willes (1963) mentioned that, although this species is gregarious, it is difficult to count because the birds are often found in rough water far from shore.

Pair-forming behavior. McKinney (1959) observed pair-forming behavior of black scoters in Alaska during April and May, and Bengtson (1966) described comparable display patterns that he observed among common scoters in Iceland during May and June. It is possible that important species variations in these behavior patterns exist, such as occurs in call duration differences among males (Sangster, 2009). However, McKinney's and Myres's (1959a) observations on the American race closely agree with those of Bengtson and myself (1965) on postural displays of common scoters.

Social courtship usually occurs in small flocks that typically contain a single female and 5 to 8 males, in McKinney's observations. Bengtson noted that as the spring progressed, the number of males in courting groups with single females increased from an average of about 4 in late May to more than 10 in late June, no doubt reflecting the gradual reduction of available females.

Bengtson found that paired males performed many of the same postures as those seen in courting groups, but in markedly different relative frequencies. Paired males exhibited the highest incidence of lateral head-shaking, general shaking (upward-stretch), and wing-flapping, while the incidences of the body-up (neck-stretching of Myres), tail-snap, low-rush, short flight, and steaming were all slightly or distinctly more frequent among males in courting parties. Preening movements appeared to be most frequent in nonaggressive unpaired and paired males and least frequent in aggressive unpaired males. Definite inciting behavior by females has not been described, although threat-like bill-pointing movements have been seen (McKinney), as well as slight chin-lifting movements (Myres).

Copulatory behavior. To judge from the few available observations, the precopulatory behavior of black and common scoters is simple and usually very short. The female seemingly adopts a prone position after both sexes have performed preening movements in various places. The male then typically performs a shake (upward-stretch) and mounts immediately. After treading, he usually swims away from the hen in a neck-stretching posture, while uttering his typical whistled notes. Some variations in postcopulatory behavior have been reported (Johnsgard, 1965).

Nesting and brooding behavior. Bengtson (1970) noted that the female common scoter "sits very tight" during incubation and is normally abandoned by the drake shortly after incubation begins, with some males remaining in the vicinity of the nest for as long as a week. Males typically then move out of the nesting areas and migrate to traditional molting areas. The limited brood counts that are available suggest that brood mergers, as often occur in white-winged scoters and eiders, are also found in this species after females abandon their broods while they are flightless, leading to brood amalgamations (Bondage and Savard, 1995).

Postbreeding behavior and populations. Postbreeding movements of the North American population of black scoters are still little understood. A major molting area for the eastern population is Hudson Bay; up to 90,000 molting black scoters have been reported from western and eastern coastal areas of James and Hudson Bays in Ontario and Manitoba, which perhaps included some scoters other than black scoters (Bordage and Savard, 1995). In western North America some of the described molting sites include the Alaska Peninsula coastline, the Yukon-Kuskokwim Delta, around Hazen Bay, Cape Vancouver, and Cape Avinof (Baldassarre, 2014).

Molt migrations are certainly present in common scoters. In Great Britain, fairly large flocks of molting male common scoters might be seen in late summer, while the largest flocks of wintering birds include females and young birds as well. Salomonsen (1968) described the molt migration of common scoters to the west coast of Jutland, in the North Sea. There, up to 150,000 birds congregate in August and September, in waters less than 10 meters deep. Birds from much of the Scandinavian and north Russian breeding populations also occur there and probably constitute the majority of these populations. These include immatures, which might arrive there in spring, as well as adult males and possibly also some nonbreeding or unsuccessful females.

Black scoters are minor game birds in North America, with about 85 percent of the total US kill shot in the Atlantic Flyway, and the annual total generally 10,000 to 20,000 birds. The Canadian kill averages smaller, with the majority taken in Quebec (Bordage and Savard, 1995). All told, during the early 2000s there might have been about 150,000 black scoters in the western North American population, 200,000 in the eastern population, and a global population of 2.1 million to 2.4 million combined black and common scoters (Baldassarre, 2014).

Numbers of all three scoter species in North America have declined greatly over the long term; breeding bird surveys indicated a total of 1.05 million scoters in 2012 over the entire survey route of Canada and Alaska, an approximate 50 percent decline from the highest numbers of more than 2 million estimated during the 1950s (Canadian Wildlife Service Waterfowl Committee, 2013).

Surf Scoter
Melanitta perspicillata (Linnaeus) 1758

Other vernacular names. Coot, skunk-head coot

Range. Breeds in North America from western Alaska eastward through Canada's Yukon and Northwest Territories to southern Hudson Bay, and in the interior of Quebec and Labrador. Winters on the Pacific coast from the Aleutian Islands south to the Gulf of California, and on the Atlantic coast from the Bay of Fundy south to Florida, with smaller numbers in the interior, especially on the Great Lakes.

Subspecies. None recognized.

Measurements. *Folded wing:* Palmer (1976): 12 males 233–252 mm, ave. 245 mm; 12 females 215–240 mm, ave. 228 mm. Delacour (1959): males 240–256 mm; females 223–235 mm.
　Culmen: Palmer (1976): 12 males 36–40 mm, ave. 37 mm; 12 females 37–40 mm, ave. 38.8 mm. Delacour (1959): males 34–38 mm; females 33–37 mm.

Weights (mass). Nelson and Martin (1953): Ave. of 12 males 2.2 lb. (997 g), max. 2.5 lb. (1,133 g); ave. of 10 females 2.0 lb. (907 g), max. 2.5 lb. (1,133 g). Vermeer and Byrd (1984): Ave. of 64 males 1,153 g; ave. of 26 females 1,025 g. Savard, Bordage, and Reed (1998): Ave. of 31 males 1,059 g; ave. of 29 females 859 g.

Identification

In the hand. Obviously a diving duck, on the basis of its enlarged hind toe and the outer toe as long or longer than the middle toe. Specimens can be identified as surf scoters if the outermost primary is longer than the adjacent one, and feathering extends forward on the culmen almost to the rear edge of the nostrils. Additionally, there is an unfeathered rounded or squarish black mark on the side of the bill near its base. Surf scoters have a maximum wing measurement of 256 mm in males (minimum 269 mm in white-wings) and 240 mm in females (minimum 261 mm in white-wings). Wing measurements overlap between black and surf scoters, but surf scoters have slightly shorter culmens (maximum 40 mm) than do black scoters (minimum 40 mm), and the differences in bill feathering noted above are definitive.

In the field. A maritime species that sometimes is found on large lakes or deep rivers during fall and winter, surf scoters can be distinguished on the water by the white markings on the male's forehead and nape, and the whitish cheek, ear, and nape markings of females. The white eye and eye markings of adult males is often visible from a considerable distance, but both sexes lack white on the wings. When landing, displaying males frequently hold their wings upward and skid to a stop in the water, and when swimming they usually hold the level of the bill slightly below horizontal. The male reportedly has a liquid, gurgling call

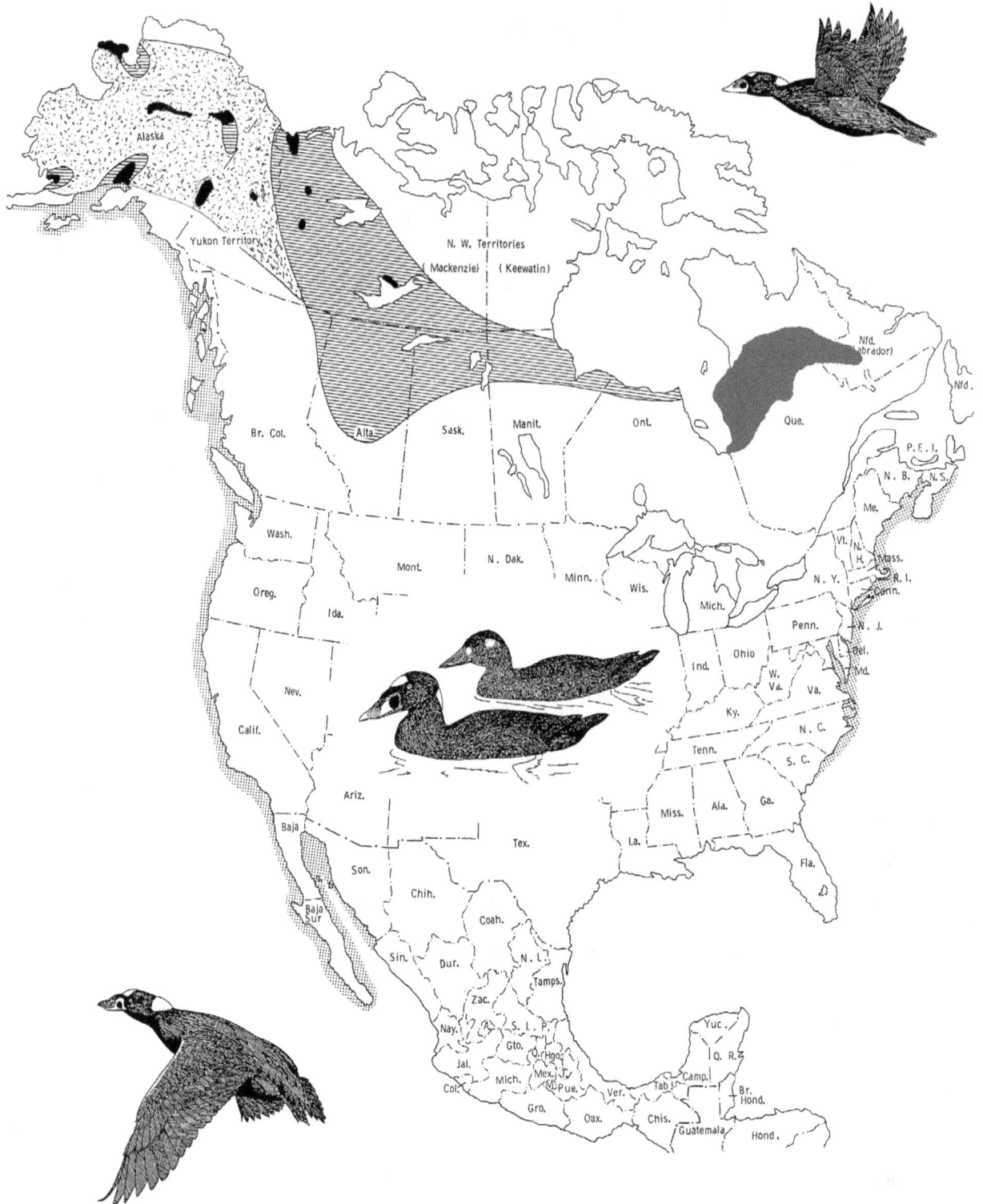

The North American known breeding (inked), probable breeding (hatched), possible breeding (stippled), and wintering (shaded) ranges of the surf scoter. Mostly after Savard, Bordage, and Reed, 1998.

uttered during courtship, and the female has a more crow-like note. In flight, their wings produce a humming sound, and like other scoters the birds usually fly in irregular lines fairly close to the water.

Age and Sex Criteria

Sex determination. The presence of black feathers anywhere on the body is indicative of a male but might not serve to separate all first-year males from females, at least prior to October, when the first blackish feathers begin to appear on the head, scapulars, and flanks of first-year birds (Bent, 1925). The males' eyes change from brown to yellow during the winter, and later to white by the end of the first year (Dwight, 1914).

Age determination. First-year females can probably be distinguished from older ones by their more conspicuous white patches on the lores and ear region. Males less than one year old lack the white forehead patch and have a less colorful bill than full adults. Their iris color might still be somewhat brownish in some yearlings. Adult plumage changes are not well studied in scoters, but there is reduction of the white forehead markings among adult males during late summer or fall (Bent, 1925), leaving only the white nape. Some adult females also develop a male-like whitish nape patch, but there is much individual variation in this (Dwight, 1914). Iverson, Esler, and Boyd (2003) provided additional information on ageing surf scoters using plumage traits.

Distribution and Habitat

Breeding distribution and habitat. In contrast to the other two scoter species, the surf scoter is entirely limited to North America as a breeding bird. This is rather surprising, in view of its widespread occurrence on this continent, and its marine wintering distribution. Its failure to colonize eastern Asia is thus difficult to understand.

In Alaska, surf scoter are widespread in summer, but many appear to be nonbreeding birds. Positive breeding records are mostly from the Bering Sea and Arctic Ocean coasts and from the upper Yukon Valley. Clutches have been found or ducklings seen at Lynx Lake (north of Bristol Bay), Kotzebue Sound, Denali National Park, Fort Yukon, and on the Porcupine River (Gabrielson and Lincoln, 1959). Presumptive breeding places in Alaska reported by Stehn et al. (2006) on aerial surveys over various years included the Tanana–Kuskokwim Valleys (2,731 birds), Koyukuk/Kanuti Valleys (1,824), Selawik–Noatak Valleys (1,797), Innoko Valley (1,363), Bristol Bay (1,360), Yukon–Kuskokwim Delta (882), Kenai–Sustna Valleys (849), Yukon Flats (552), and the Arctic Coastal Plain (222).

In Canada, the species likewise is of widespread occurrence, although few breeding records exist. In British Columbia there were only 7 confirmed breeding records for the province prior to 1990, all in the Peace River District. Perhaps fewer than 100 pairs breed in the province (Campbell et al., 1990; Davidson et al., 2008–2012). Godfrey (1966) included within the breeding range Yukon Territory (probably); western McKenzie District, Alberta (Elk Island Park); northern Saskatchewan (Lake Athabasca); James Bay (Charlton and Sheppard Islands); northern Ontario; Quebec (Wakuach Lake and near Otelnuk Lake); and Labrador (Grand Falls and Petitsikapau Lake).

More recent studies (e.g., Reed, Aubry, and Reed, 1994; Savard, Bordage, and Reed, 1998; Morrier et al., 2007; Lesage, Reed, and Savard, 2008) suggest that Quebec is a major breeding region for the species, with relatively high densities at Laforge Reservoir, Nastapoka River, and Lake Bienville. Breeding also occurs in central Labrador, although surveys there have not revealed large numbers (Baldassarre, 2014). Breeding habitat requirements are little known but probably are like those of the other scoters. Thus, freshwater ponds, lakes, or rivers, with shrubby cover or woodlands for nesting cover in the vicinity, are probably typical.

Wintering distribution and habitat. The wintering distribution of this species is much better known than its breeding distribution. The overwhelming majority of birds evidently winter along North American coastlines, although stragglers do occur along the coastlines of Asia and Scandinavia. Surf scoters winter abundantly in the waters of southeastern Alaska, especially along the Alexander Archipelago, with lesser numbers extending along the south coast of the Alaska Peninsula and to some extent into the Aleutian Islands (Gabrielson and Lincoln, 1959).

In western Canada, about a half million surf scoters winter along the coast of British Columbia, and they, together with white-winged scoters, are perhaps the most numerous of all the wintering ducks of the Puget Sound region of Washington (Jewett et al., 1953; Davidson et al., 2008–2012). Phillips (1926) suggested that from Puget Sound northward, surf scoters tend to be outnumbered by white-winged scoters, whereas farther south the reverse presumably applies. Aerial counts in the late 1950s and early 1960s indicate that scoters then constituted nearly half of the wintering diving ducks of this area (Wick and Jeffrey, 1966). Surf scoters also winter commonly along the Pacific coastlines of Oregon and California and are the commonest scoter species of northwestern Mexico. Leopold (1959) noted more than 24,000 during winter inventory counts, with the largest population in San Ignacio Bay of Baja Peninsula.

On the Atlantic coast, surf scoters winter from Newfoundland and the Gulf of St. Lawrence southward (Godfrey, 1986), with occasional birds appearing on Lakes Erie and Ontario. Atlantic coast populations are seemingly not so large as those on the Pacific coast. In the Chesapeake Bay area the surf scoters are usually the commonest scoter species in the coastal sections, while white-winged scoters are much more common in the Bay proper (Stewart, 1962). As far south as South Carolina the surf scoter is still a fairly common winter visitor (Sprunt and Chamberlain, 1949). Preferred wintering habitats include the littoral zone of the ocean and adjoining coastal bays, with a few birds utilizing salt or brackish estuarine bays in the Chesapeake region (Stewart, 1962).

General Biology

Age at maturity. Age at maturity is not definitely established, but judging from their molting sequence the birds probably breed at the end of their second year. A fully mature plumage and bill coloration might not be attained until the following fall (Bent, 1925).

Surf scoter, male in breeding plumage

Pair-bond pattern. Pairs are apparently reestablished each winter and spring, during a prolonged period of social display (Myres, 1959a).

Nest location. Too few nests have been found to allow many conclusions on nest selection. McFarlane (quoted by Bent, 1925) reported that the nests are usually located at a considerable distance from water and always well concealed under the low-spreading branches of a pine or spruce tree. Nests are typically placed in dense shrubs or under overhanging conifers, or under fallen logs well away from shore, with a preference for nesting on lake islands (Kear, 2005).

Bent noted that in Labrador the ducks reportedly nest about the inland ponds and lakes, placing their nests in grass or under bushes close to the edge of the water. Morrier et al. (2001) reported that the birds disperse widely while nesting, with most small lakes supporting only a single pair, although Reed, Aubry, and Reed (1994) noted that a relatively large (640-hectare) lake supported three pairs.

Clutch size. Probably from 5 to 7 eggs normally constitute a clutch, with larger numbers unusual (Bent, 1925). Morrier et al. (2007a) reported that the average of 18 Quebec clutches was 7.6 eggs, including 1 clutch of 6 eggs, 7 of 7 eggs, 8 of 8 eggs, and 2 of 9 eggs.

Incubation and fledging period. The incubation period is probably 28 to 30 days (Savard et al., 1998b), and the egg-laying interval is about 1.5 days per egg. Ducklings are fully feathered by 42 to 55 days, and are abandoned by most parents by that age (Lesage et al., 1997). Fledging occurs at about 55 days of age.

Nest and egg losses. Little specific information is available on egg and nest losses, but over two years of study at Lake Malbaie, Quebec, 12 of 17 nests hatched (Morrier et al., 2007a). Eight of 17 nests hatched near Laforge Reservoir, Quebec, and over two years of study 57 broods were seen in a population collectively totaling 122 pairs (Lesage et al., 1997), suggesting a hatching success of about 46 percent. Brood amalgamation is fairly common, with a single female often tending such merged broods (Savard et al., 1998b).

Juvenile and adult mortality. Little information is available regarding mortality; immature and adult survival rates in the wild are still unreported. Lesage et al. (1997) reported that over two years 47 and 65 pairs were present on Lake Malbaie, and that a total of at least 421 ducklings were produced during the two years, an average of 3.76 ducklings per brood. There was an estimated 65 percent duckling mortality rate in the first year, and 55 percent in the second year, resulting in breeding success rates of 0.85 and 2.14 fledglings raised per pair during each of the two years. The high mortality rate during the earlier year was attributed to unusually cool and wet early summer conditions, while the high rate the second year might have been the result of little or no gull predation.

One indirect estimate of first-year survival in surf scoters was provided by Iverson et al. (2004a), who used age ratios as an index of recruitment, and who found that first-year males composed 7 to 13 percent of wintering males, and first-year females 23 percent of all females. No estimates of typical survival rates of older birds in the wild are available, but a female banded as an adult survived for nearly eight years (Baldassarre, 2014), and one male lived for nearly ten years at the San Diego Zoo (Delacour, 1959). Probably relatively few surf scoters are killed by hunters, as they are not heavily hunted in most regions. Surf scoters constitute only about one percent of the waterfowl shot in the Atlantic Flyway, and about 0.2 percent of the national total (Savard, Bordage, and Reed, 1998).

General Ecology

Food and foraging. Cottam's (1939) analysis of food samples from 168 adult surf scoters taken throughout the year is a major source of information about scoter food and foraging. He reported that mollusks (especially blue mussels and related mollusks) constituted 60 percent of the food volume, with crustaceans and insects another 10 percent each, and various plant materials totaling about 12 percent. As with the other two scoters, bivalve mollusks, especially blue mussels (*Mytilis edulis*), make up at least half of the surf scoter's foods; clams, oysters, and scallops apparently are less frequently utilized. Cottam also reported that 7 juvenile birds had fed largely on various insects and, to a lesser extent, on mollusks and freshwater or terrestrial plants.

Cottam judged that most foraging was done in early morning, since many birds shot during midmorning hours already had empty stomachs. The birds often forage in water just beyond the breakers, usually in depths from 6 to 30 feet. During observations in Vancouver harbor, I noted that the surf scoters were foraging in shallower waters and closer to shore than were the much less common white-winged scoters. However, they apparently regularly are associated with that species in wintering areas, even though white-winged

Surf scoter, male in flight

scoters seemingly depend to a greater extent on oysters, clams, periwinkles, and mollusks other than mussels (Cottam, 1939; McGilvrey, 1967).

Among gizzards analyzed from 133 surf scoters wintering along Massachusetts and New Hampshire, 60 percent of the animal foods were of wedge clams (*Mesodesma*), 24 percent were Atlantic razor clams (*Ensis*), and 8 percent were blue mussels (Stott and Olson, 1974). Clams are important foods for all three species of scoters in that region, and require sandy substrates. However, Anderson, Lovvorn, and Wilson (2008)

Surf scoter, male (alert posture)

noted that traditional methods of estimating sea duck foods (e.g., from gizzard contents) have greatly underestimated the importance of soft-bodied invertebrates in the diet of surf scoters, such as polychaetes, and have overestimated the importance of hard-shelled invertebrates.

In a six-study comparison of foods found in the gizzards of surf scoters, among mollusks, blue mussels were present in all six; rock clams (*Prototheca*) were in five; and razor clams, wedge clams, manila clams, and rock clams were in one each. Among crustaceans, barnacles were in all six, and other crustaceans were present in three. Polychaete annelids, insects, herring eggs, and plants were found in small quantities (Savard, Borage, and Reed, 1996).

Sociality, densities, territoriality. No specific information is available about surf scoter sociality, densities, and territoriality. The few available observations indicate that nests are widely scattered over large areas (Reed, Aubry, and Reed, 1994; Morrier et al., 2004) and that typical territoriality is generally absent in scoters, with the male defending only his mate.

Surf scoter, male (relaxed posture)

Interspecific relationships. All three species of scoters utilize much the same habitats where they occur together, and perhaps some foraging competition does exist among them. Surf scoters apparently are closest to the black scoter in the kinds of foods utilized, but surf scoters seemingly winter in more southerly locations and forage closer to the coasts than other scoters. Cottam noted that they eat fewer of the heavier-shelled mollusks than do the larger scoter species, and also are possibly more partial to foods of vegetable origin. Too little is known of the nesting biology of surf scoters to judge the possible importance of nest and duckling predators (but common loons are said to cause panic among broods and have been reported to kill ducklings), or to judge other important interspecific relationships occurring at this highly vulnerable stage in the life cycle.

General activity patterns and movements. Jewett et al. (1953) reported that the surf scoter is extremely active during the morning and evening hours, coming inshore as far as it possibly can and diving for food in the shallows, where animal life is the most abundant. Like the other scoters, it probably retires to deeper waters to spend the night, although it is possible that some nocturnal foraging activity occurs under favorable conditions.

Social and Sexual Behavior

Flocking behavior. During migration and while on wintering areas, flocks of hundreds or even thousands of scoters are not uncommon and might be of single-species groups or of mixed composition. In the Puget Sound area, white-winged and surf scoters occur in flocks of 50 to 2,000 or more birds, the two species often about equal in numbers. Large flocks have been recorded until about April, but by May groups of 40 to 50 are more usual. During the northern movement in spring, migrant flocks are often of larger size and tend to gather in compact rafts (Jewett et al., 1953).

Pair-forming behavior. Most of the available information on surf scoter pair-formation activities derives from the work of Myres (1959a) in the Vancouver area. During late winter and spring, social display might be easily seen there. In late March I observed several small groups of courting birds displaying simultaneously, while the majority of the visible birds were apparently already paired and were engaged in foraging behavior. A good deal of overt or ritualized threatening behavior was evident in these groups, with the males often attacking one another and with the female threatening any male that approached too closely. Myres mentioned observing the females performing chin-lifting movements and uttering a crow-like note. On the basis of my limited observations, I regard this as functionally equivalent to the inciting behavior of other sea ducks.

Male postures and movements are several, including aggressive crouched and threat postures much like those of male goldeneyes. A common male posture is the "sentinel," in which the bird vertically stretches his neck to the utmost, with the tail either raised vertically or in the water. From this posture the male might begin "breast-scooping," which appears to be a ritualized version of breast-preening movements. A liquid, gurgling call accompanies these movements. A short flight, or "fly-away" display, is also common, and on landing the male holds his wings in an upward V posture as he skids to a stop in the water. Probably the most obviously ritualized display is "chest-lifting," a sudden and energetic vertical chest-lifting movement, usually performed close to the female and seemingly directed toward her. I did not hear any calls associated with this display, and was greatly impressed by its similarity in form and apparent function to the "rearing" display of the male Steller's eider, which is also silently performed.

Copulatory behavior. Myres (1959a, 1959b) has provided the only detailed observations on copulation in the surf scoter. He observed four instances between late December and early January, and in no case did the birds appear to be permanently paired. The female assumed a prone posture and remained in it for some time, in one case for up to about 2 minutes. The male performed water-twitching (dipping and shaking the bill in the water), preening-behind-the-wing, and also "false" (ritualized) drinking. During treading the male flicked his wings, and on each of the four occasions he performed a single chest-lifting display as he released his grip on the female. No other specific postcopulatory displays by either sex were noted.

Nesting and brooding behavior. Few studies on the nesting behavior of this species have been performed, but Morrier et al. (2007) described 17 nests found on islands in Lake Malbaie, Quebec. The nest placements averaged 19 meters from shore, and all of them were all located in standing forest vegetation or among windfalls. They were all well hidden and completely obscured when viewed from above, with fallen tree trunks often providing overhead visual cover.

Postbreeding behavior. Many immature birds spend their summers along the Pacific coastline and especially along coastal Alaska. Gabrielson and Lincoln (1959) noted that summer flocks sometimes occur along the south shore of the Alaska Peninsula and around some of the Aleutians (Gibson and Byrd, 2007). They reportedly become plentiful at Sitka about August 15. Known major molting sites in western North America include the Mackenzie Delta, Alaska, and coastal British Columbia (Kear, 2005).

On the Pacific coast, up to 8,000 male surf scoters pass Alaska's Cape Pierce during July on their way to molting areas; females might arrive at molting sites later than males, since they begin their wing molts about 3 to 4 weeks later than the males. Some important Alaskan molting areas include the Yukon Delta coast and, in British Columbia, Boundary Bay, near Prince Rupert, the Queen Charlotte Islands, and Howe Sound (Baldassarre, 2014).

In the interior and Atlantic coast some major molting centers reported include Hudson Bay and northeastern James Bay, the St. Lawrence estuary (with up to 50,000 surf and white-winged scoters), and coastal Labrador (50,000–62,000 scoters, mostly male surf scoters).

White-winged Scoter
Melanitta fusca (Linnaeus) 1758

Other vernacular names. Velvet scoter (Britain), white-winged coot

Range. Breeds (*M. f. deglandi*) in North America from northwestern Alaska east from Yukon Territory, Northwest Territories, and southwestern Nunavut to Hudson Bay, and south through western Canada to southern British Columbia, southern Alberta, southern Saskatchewan, and southern Manitoba; previously (now rarely) south to north-central North Dakota. Winters on both coasts of North America, from Alaska to Baja California, and from the Gulf of St. Lawrence to South Carolina. *M. f. fusca* breeds in Scandinavia, Estonia, northern Russia, and northeastern Siberia, wintering southward coastally.

North American subspecies. *M. f. deglandi* (Bonaparte). American White-winged Scoter. Breeds in North America as indicated above. Treated taxonomically as a species separate from *M. fusca* by Livezey (1995) and some European authorities.

   M. f. fusca (L.). Velvet Scoter. Breeds in Europe and Asia; in North America. Migrants occur casually in Greenland.

Measurements. *Folded wing:* Delacour (1959): Males 269–293 mm, females 251–266 mm, Palmer (1976): males 274–298 mm, females 256–280 mm.

   Culmen: Delacour (1959): Males 37–50 mm, females 38–43 mm. Palmer (1976): males 36–47 mm, females 35–43 mm.

Weights (mass). Nelson and Martin (1953): 13 *M. f. deglandi* males ave. 3.4 lb. (1,542 g), max. 4 lb. (1,814 g); 19 females ave. 2.7 lb. (1,223 g), max. 3.4 lb. (1,542 g). Schiøler (1926): 9 older *M. f. fusca* males ave. 1,727 g, max. 2,104 g; 6 juveniles *M. f. fusca* ave. 1,670 g; 11 older *M. f. fusca* females ave. 1,658 g; 4 juveniles ave. 1,214 g.

Identification

In the hand. As with other diving ducks, the enlarged hind toe and lengthened outer toe is present, and specimens can be recognized as a scoter by the heavy bill and rather uniformly dark body. Unlike the other scoters, it has a bill that is feathered forward to a point near the posterior edge of the nostrils, its outermost primary is not appreciably narrower than the adjoining one, and the wing speculum on the secondary feathers is white. The white-winged is the largest of the scoters, with its folded wing measuring at least 269 mm in males and 251 mm in females; surf scoters have a maximum wing of 256 mm in males and 235 mm in females. Among male European and Asian velvet scoters (*M. f. fusca*) the bill is less swollen basally and is

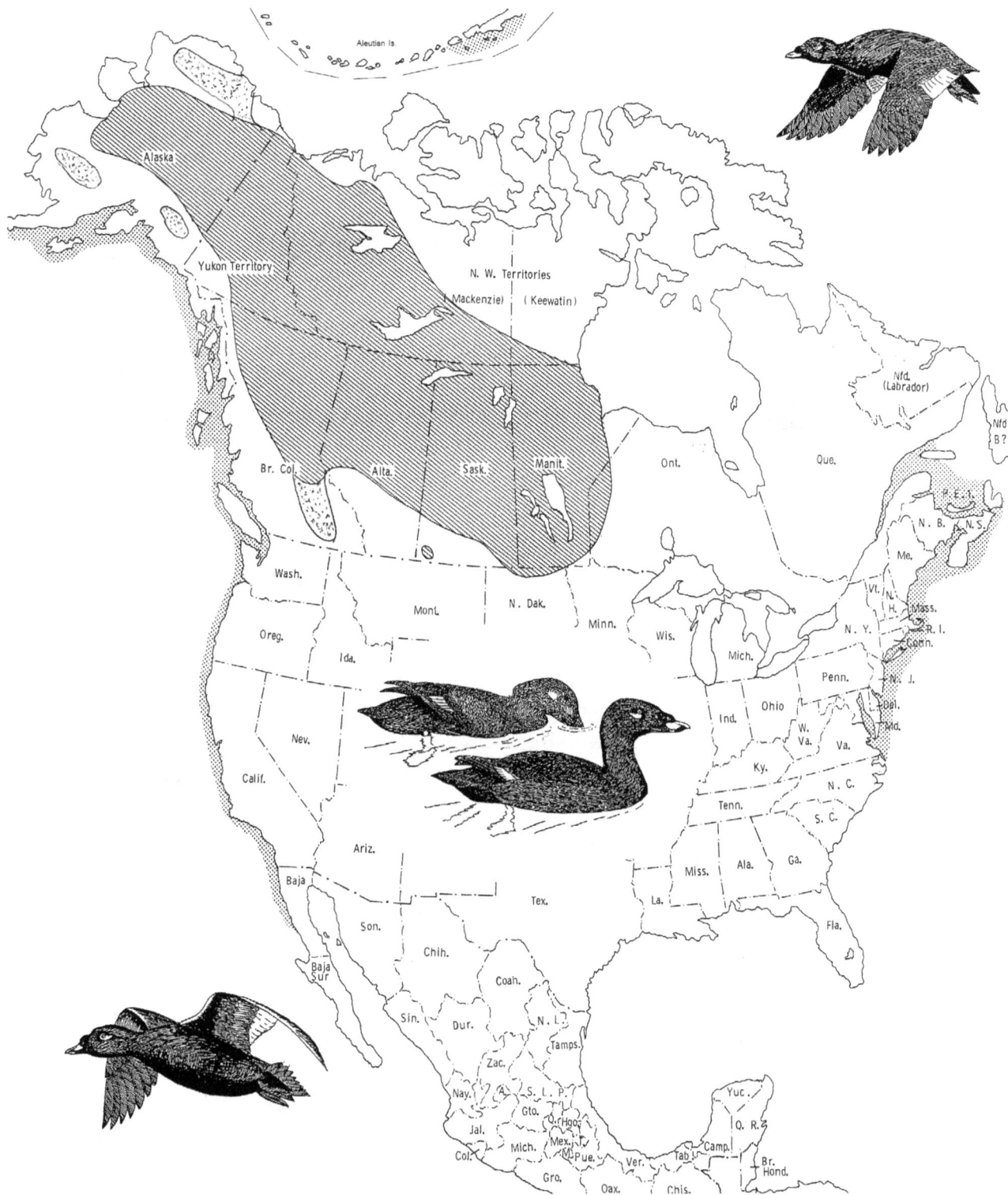

The North American breeding (hatched) and wintering (shaded) ranges of the white-winged scoter. Regions of newly acquired or previously unrecognized breeding are shown by stippling.

mostly yellow, whereas in the American form (*M. f. deglandi*) the bill of adult males is more much swollen and the sides are mostly orange to red.

In the field. White-winged scoters are usually found on the coast but are more likely than the other two scoters to also be seen on large interior lakes during winter. On the water the white wing markings are sometimes not visible, and a white eye and eye-patch on the male might be the only apparent parts of the bird that are not dark brown or black. Adult females closely resemble female surf scoters on the water but never exhibit whitish nape markings. The blackish crown of the white-winged female also contrasts less sharply with the sides of the head, and the pale cheek and ear markings are generally less evident than in the surf scoter. When the birds flap their wings or fly, the white secondaries provide the best field marks. In flight, white-winged scoters are the most ponderous of the scoters, usually flying low over the water in loose flocks or long lines and producing very loud wing noises (Bent, 1925), although their outermost primaries are not noticeably narrowed. Vocalizations in both sexes seem to be weak, judging from recent descriptions.

Age and Sex Criteria

Sex determination. By December or later, first-year males begin to acquire the black feathers by which all older males can be readily separated from females, with the first such feathers appearing on the head. For younger birds, internal examination is necessary to determine sex.

Age determination. Males less than one year old have brownish underparts and a less colorful and swollen bill than do adults, and until 14 to 15 months of age have a brown iris (Bent, 1925). A fully black body and wing plumage color is attained by males during the second winter, but maximal bill size is evidently not developed until about their third year. First-year females have more conspicuous whitish markings on the cheeks and ear region than do older females, which might exhibit almost no pale marks on the sides of the head (Dwight, 1914). Some young females have a very much reduced white speculum, and additionally the outer side of the tarsus is blackish, while the inner side of the tarsus and toes are dull purplish brown, rather than yellow to brownish red (Kortright, 1943). Immatures of both sexes exhibit light, frayed tips on the tertials and tertial coverts, and the greater coverts often are entirely brown, or have less white on their tips than is present in adults (Carney, 1964).

Distribution and Habitat

Breeding distribution and habitat. The North American breeding distribution of this most widespread species of scoter is almost entirely limited to Alaska and Canada. In Alaska all definite records of breeding are from the interior, chiefly in the vicinity of Fort Yukon. From this area the birds also breed eastward along the Porcupine River drainage, northward at least as far as Anaktuvuk Pass, south to the Minto Lakes, and west to the Innoko watershed and the vicinity of Koyukuk (Gabrielson and Lincoln, 1959).

White-winged scoter, male in breeding plumage

In Canada, the white-winged scoter is the commonest and most widespread breeding scoter species. It ranges from the mouth of the Mackenzie River southward through the Yukon and western Mackenzie District to central British Columbia, where probably fewer than 10,000 breed in the Boreal Plains and central interior ecoregions (Davidson et al., 2008–2012) south to the Thompson-Okanagan Plateau. It breeds widely across the prairie wetlands and forested portions of Alberta, Saskatchewan, and Manitoba, with the eastern breeding limits in the vicinity of Churchill, Manitoba, and Ney Lake, Ontario. Breeding also occurs in the Cypress Hills area of southwestern Saskatchewan and as far south as Shoal Lake, Manitoba (Godfrey, 1986). The highest densities are in northwest Canada, in the boreal forest zone between Great Slave Lake and the Arctic Ocean (Brown and Fredrickson, 1997).

South of Canada, white-winged scoters often summer in coastal areas and sometimes occur in the interior states as nonbreeders during the summer months, but apparently only in North Dakota has any breeding occurred. In the early 1900s the birds regularly bred in the vicinity of Devils Lake, but they apparently became rare in North Dakota by 1920. Since then, a few broods have been seen in McHenry County, and at Des Lacs and Lostwood refuges, Burke County (Duebbert, 1961).

Habitat requirements of white-winged scoters have not been well analyzed in North America, but studies on the European race probably are applicable to this region. There, nesting on open tundra is rare, and the coastal archipelagos and lakes of the northern coniferous forest zone seem to represent the original breeding habitats of the species. Hildén (1964) found nesting to occur in open scrub heaths and birch woods of larger islands of the Valassaaret group in Finland, as well as on small islets.

Boulder islets dominated by herbaceous vegetation and with shrubs and trees present seem to represent the ideal habitat type and provide suitable shrubby nesting cover and often associated gull nesting colonies, to which this species is attracted. Islets richly overgrown with bushes or partially wooded sometimes have scoters nesting even in the absence of gulls, indicating that these vegetational features are an important aspect of breeding habitat selection. Unlike common eiders of the same area, the scoters nest all the way to the central parts of the larger islands. Favored brood habitats are those with extensive shoals and shallow, narrow water areas sheltered from heavy wave action.

Wintering distribution and habitat. In Alaska, the bays and channels of the Alexander Archipelago of southeastern Alaska seem to be the center of abundance of wintering white-winged scoters, but the species extends in smaller numbers westward to the eastern Aleutian Islands and is regularly seen around the Pribilof Islands (Gabrielson and Lincoln, 1959).

In western Canada, this species winters commonly along the coastline of British Columbia (Godfrey, 1986) and together with the surf scoter is among the commonest wintering ducks of the Washington coast and Puget Sound area (Jewett et al., 1953). It is a common to abundant winter resident along the coasts of Oregon and California and is of regular occurrence as far south as San Quintín Bay (Leopold, 1959).

In eastern Canada, white-winged scoters winter from Newfoundland and the Gulf of St. Lawrence southward to the United States border and, in limited numbers, also on the Great Lakes (Godfrey, 1986). From Maine southward they are relatively abundant along the New England coastline. Like the other two scoters, they prefer the littoral zone of the ocean for wintering habitat, just beyond the breakers and within a mile of shore. In the Chesapeake Bay area they are the commonest of the three scoter species (Stewart, 1962).

In the Chesapeake Bay area this species is common in coastal bays and in salt or brackish estuarine bays, with some birds extending into slightly brackish estuarine bays (Stewart, 1962), but farther south in South Carolina they are the least common of the scoters (Sprunt and Chamberlain, 1949). White-winged scoters are rare in Florida and Louisiana. They are also rare through most of the interior of the United States south of Canada, although stragglers sometimes winter on reservoirs and rivers.

General Biology

Age at maturity. Sexual maturity reportedly occurs in the second year of life (Dementiev and Gladkov, 1967). Females are known to breed at as early as two years of age, but some might not breed until their third year (Brown and Houston, 1997).

Pair-bond patterns. Pair-bonds are normally renewed each year, with pair-forming behavior beginning on wintering areas. Males desert their mates during early stages of incubation and make sometimes extensive flights to molting areas. Rawls (1949) noted that males began to band together at Delta, Manitoba, in mid-July, or about a month before the first broods were seen. This would indicate a desertion of females at the time incubation begins.

White-winged scoter, male in flight

Nest location. Hildén (1964) reported on 254 nest sites he observed on the Valassaaret Islands of the Gulf of Bothnia. About three-fourths of these were well concealed, in most instances underneath junipers (29.4 percent) or *Hippophae* bushes (24.2 percent). Forb cover accounted for another 17.1 percent, and the remainder were in mixed shrub-forb cover, other cover types, under boulders, or were exposed. Exposed nests were typical only on islets having a moderate to high density of nesting gulls.

Koskimies and Routamo (1953) reported an even higher usage of juniper bushes (82 percent) on the relatively large and partly wooded islands of their study area. The height and density of the surrounding bushes are apparently not important, but the nests are most often placed under stones projecting from the earth, in a cavity, or among stones that are well covered by juniper branches and stems. The shoreline zone is generally avoided for nesting, even though dense grasses sometimes occur there. Instead, nests are generally placed among woodland far from shore, and on small islets lacking such cover the species does not nest at all (Hildén, 1964). Relatively few nests from the North American population have been described, but roses (*Rosa*), willows (*Salix*), raspberries (*Rubus*), and gooseberries (*Ribes*) are apparently favored shrubby cover (Rawls, 1949).

Clutch size. Hildén (1964) reported an average clutch size of 8.43 eggs for 187 clutches, with a modal clutch of 9 eggs and a range from 6 to 12. Koskimies and Routamo (1953) also found an overall average of 8.43 for 90 Finnish clutches, with earlier clutches averaging slightly larger than later ones. Comparably sized samples from North America are those of Brown and Brown (1981) who reported a mean clutch of 9.17 eggs for 146 clutches from Alberta and Saskatchewan, and Brown and Fredrickson (1989), who later reported a mean of 8.85 eggs for 426 clutches. Vermeer (1969) noted an average clutch size of 10.2 eggs in 12 Alberta clutches, and a range of 6 to 16 eggs. The average egg-laying interval is about 40 hours.

Incubation period. The incubation period for 21 Alberta nests averaged 27.7 days but ranged from 25 to 30 days (Brown, 1977). The average of 29 nests was 27.2 days for birds of the European race, with an observed range of 26 to 29 days (Paavolainen, cited in Bauer and Glutz, 1969).

Fledging period. Fledging occurs at 70 to 80 days of age. At this time captive-raised birds had a body mass of about 60 to 70 percent that of adults; adult mass was not achieved until about 21 weeks of age (Brown and Fredrickson, 1983, 1997). This is the longest period of development for any North American duck species (Alisaukas et al., 2004) and might represent a serious disadvantage when breeding at high latitudes having short breeding seasons.

Nest and egg losses. Nest success in Saskatchewan (Redberry Lake) and Alberta (Jessie Lake) studies varied from 65 percent (49 nests) to 76 percent (99 nests), with a larger sample of 432 nests averaging 73.1 percent (Brown and Brown, 1981; Brown and Fredrickson, 1989). About half of all nests these researchers found were deserted, and most nest destruction was caused by gulls and crows, but hatching success in successful nests ranged from 67 to 92 percent. A lower hatching success of 29.5 percent was reported for Redberry Lake by Traylor, Alisauskas, and Kehoe (2004a), with both avian and mammalian predators being significant mortality factors.

Hildén (1964) provided hatching data over two years of study for Finnish velvet scoter nests, which had a remarkably high overall average nesting success of 91 percent. Compared with common eider nests on the same study area, the rate of scoter nesting success was higher and seemingly attributable to the species' more sheltered nest locations and a lesser tendency toward nest desertion. Evidently most egg losses to crows and ravens occur during the egg-laying period, when the eggs are poorly covered and not defended by the female. Even so, the total egg losses (15.7 percent) found for this species were fewer than those of the other ducks nesting in the area. Koskimies and Routamo (1953) likewise found a high loss to crows during the egg-laying period, but no cases of predation were found after incubation had begun.

Juvenile mortality. Because of the strong tendencies of this species to form mixed broods and for the females to abandon their young early, brood size counts provide no reliable estimate of prefledging mortality. At least in marine environments such as the Gulf of Finland, scoter duckling mortality is often extremely high, and usually exceeds 90 percent (Koskimies, 1955). Hildén (1964) likewise reported comparable velvet scoter brood losses in the Gulf of Bothnia during three years of study. During these years the loss of individual young was estimated at 92 to 100 percent, and the rate of brood losses at 83 to 99 percent. These losses seemed to be related to the low tolerance of scoter ducklings to severe weather conditions.

Apparently this species is basically adapted to breeding on relatively small inland waters, and only during years of unusually fine weather is brood success high in marine habitats. In local areas of reed bays, where the water temperature is fairly high and there is protection against rough seas, survival might be fairly high, although such sheltered areas might develop overly congested brood populations. Losses of ducklings to gulls seem to be related to variations in weather, with predation rates much higher during

White-winged scoter, male

bad weather. Herring gulls and great black-backed gulls are evidently by far the worst of these predators (Hildén, 1964).

Rawls's (1949) studies at Delta, Manitoba, indicated that brood mergers are not typical in this sparse population. Twelve broods up to 1 week of age averaged 4.75 young, four 2-week-old broods averaged 4.0 young, and three 3-week-old broods averaged 3.0 young. Brood survival in this inland location is thus seemingly higher than is characteristic of marine environments.

Adult mortality. Krementz et al. (1997) calculated an annual survival rate of 77.3 percent for adult female white-winged scoters banded at Redberry Lake. Kehoe et al. (1989) produced a similar estimate of 78.2 percent annual survival for that population. By marking female velvet scoters and observing them again on nests in later years, Koskimies (1957) calculated an annual adult mortality of only about 5 percent. Perhaps these high survival rates for adults help compensate for the often extremely low nesting and rearing success that seem to be typical of the species.

General Ecology

Food and foraging. The survey by Cottam (1939) is still the most comprehensive study of the foods of this species. He analyzed the foods found in 819 adults and 4 juveniles, most of which were taken along the coasts of Massachusetts and Washington. Among the adults, 75 percent of the foods by volume were found to be mollusks, of which rock clams (*Protothaca*), oysters (*Ostrea*), and mussels (especially *Mytilus*) were the most prevalent, and bivalves collectively constituted 63 percent of the total. Seemingly availability, rather than specific food preference, determined the types of foods taken. Crustaceans were as a group second in importance, and included various decapods (crabs, crayfishes, etc.), amphipods, and barnacles. Other foods of adults included insects, fishes, plant foods, and miscellaneous materials, all in quantities of less than 3 percent by volume. The few juveniles that Cottam examined had primarily consumed various crustaceans. Rawls (1949) noted that 4 of 5 juveniles he collected at Delta, Manitoba, had been consuming *Hyalella* amphipods, as had two adult females. *Hyalella* is evidently an important food source for this species in freshwater habitats (Brown and Fredrickson, 1986).

McGilvrey (1967), reporting on 124 white-winged scoters collected from Maine to Long Island Sound, found substantial differences in food taken according to area of collection. In birds from Maine, over half the food volume consisted of dog winkles (*Thais*). Among Massachusetts birds, blue mussels (*Mytilus*) and yoldia (*Yoldia*) constituted over 60 percent of the volume. Birds from the Long Island area had taken a wider variety of mollusks, including periwinkle (*Littorina*), yoldia, and nassa (*Nassarius*), plus a fish, the sand launce (*Ammodytes*). A similar array of mollusk foods, including blue mussels, periwinkles, whelks (*Nassa*), and cockles (*Cardium*), have been reported to be consumed by European velvet scoters by Madsen (1954).

Vermeer and Bourne (1984) reported on foods consumed in British Columbia over all four seasons. Pelecypod mollusks ranked highest in importance by wet weight measure during all seasons, with gastropod mollusks and balanomorph crustaceans roughly comparable in secondary importance.

Most observers report that white-winged scoters usually forage in water less than 25 feet deep, although dives to depths as great as 60 feet have been reported. Mackay (1891) stated that this species prefers to forage in water less than 20 feet deep but can forage in waters as deep as 40 feet. This species seems to have unusually great endurance in remaining submerged. Breckenridge (in Roberts, 1932) found that a male remained submerged an average of 57.5 seconds, with intervening average rests of 12 seconds; a female had similar average diving and resting durations of 62 and 11 seconds. Rawls (1949) reported that when adults foraged in water less than 10 feet deep, their dive durations averaged 30 seconds, with average intervening periods of 15 seconds. However, one three-day-old duckling that was being chased dove repeatedly for about 15 minutes, with each dive lasting about 30 seconds, and the periods between dives averaging only 10 seconds. Koskimies and Routamo (1958) noted maximum observed diving times of 46 and 56 seconds for females and males, respectively.

Sociality, densities, territoriality. Probably in most areas nesting densities are rather low, but on favored nesting islands the densities are sometimes considerable. Robert Smith (quoted by Rawls, 1949) found 20

White-winged scoter, male and female

nests in an area of less than half an acre on a small willow island at Chip Lake, Alberta. Comparably high nesting densities have been found in southwestern Finland, where Koskimies and Routamo (1953) observed a maximum nesting density of 9 pairs on 0.5 hectare (1.2 acres) of juniper on a small island. Hildén (1964) estimated that in 1962 there were 294 breeding pairs of velvet scoters on his refuge study area, which included 6 square kilometers of land area representing 49 pairs per square kilometer.

Territoriality is evidently lacking in this species. Rawls (1949) noted that territorial behavior seemed to be almost nil at Delta, Manitoba, and he never observed defense of any areas but did observe two cases of males defending their mates. Koskimies and Routamo (1953) also reported that after migratory flocks break up the males begin to maintain small "mated female distances," which gradually become larger as the breeding period approaches. Each pair also occupied a specific water area of varying size away from the nest site, a typical territorial criterion.

Interspecific relationships. White-winged scoters probably compete to some extent for food with surf and black scoters, since these species have very similar diets and often intermingle on wintering areas. White-winged scoters do, however, tend to winter in more northerly areas than do the smaller scoter species. Mixed clutches presumably result for similar nest site requirements; those found in Finland involved the red-breasted merganser, tufted duck (*Aythya fuligula*), and greater scaup (Hildén, 1964), whereas those in North America involved the American wigeon, gadwall, and lesser scaup (Weller, 1959). In most cases the scoters deposited eggs in the other species' nests, rather than the reverse. Dump-nesting by female scoters in nests of their own species is also fairly prevalent in areas where the birds nest in close proximity.

Crows and ravens are seemingly responsible for most of the egg losses to white-winged scoters, while various large species of gulls (herring and great black-backed, particularly) have been reported to be serious duckling predators in Europe (Hildén, 1964).

General activity patterns and movements. Like the other scoters, white-winged scoters are daytime foragers. However, they migrate either by day or by night (Cottam, 1939). Rawls (1949) indicated that a surprisingly regular daily periodicity might also occur on the breeding grounds; during seven mornings between June 26 and July 3 and within 2 minutes of 4:50 a.m. he observed a pair regularly fly from Lake Manitoba to an adjacent marshy bay, always flying over the same tree. Rawls also noted that the birds usually foraged for periods up to about 25 minutes, followed by intervening rest periods of about 30 minutes. Most of the foraging he observed on Clandeboye Bay of Lake Manitoba seemed to occur during morning and late afternoon or early evening hours. On the basis of the early morning flights that Rawls observed, the nocturnal hours were probably spent on the deeper parts of the lake.

Social and Sexual Behavior

Flocking behavior. Scoter flock sizes on wintering grounds are sometimes fairly large, especially as they congregate prior to migrations. Bent (1925) mentions noting several thousand of these birds gathered in large flocks off the coast of Rhode Island in early May. Mackay (1891) stated that such migrant flocks often number 500 to 600 birds, which typically depart during afternoon hours. On their arrival at breeding grounds, these flocks evidently break up rather rapidly into paired adults and nonbreeders.

Rawls (1949) noted that immature birds were usually seen in groups of about 5 to 30 on Lake Manitoba during the summer months but observed that most of the adults seemed to be paired on arrival. Koskimies and Routamo (1953) noted that groups of males seldom have more than 20 individuals and that summer assemblies of immatures are usually not over 30 and only exceptionally reach 60 birds.

Pair-forming behavior. Social displays have been described for the velvet scoter by Koskimies and Routamo (1953), and by Myres (1959a) for the white-winged scoter, which have not yet been closely compared for possible taxonomic differences. Primarily agonistic postures of the male in both forms include the "crouch," in which the body is low in the water and the head is tilted forward and downward at a 45-degree angle.

White-winged scoter, female

In the "alarm" posture the neck is more elongated and sloped forward. An attack or threat posture is also present and greatly resembles the corresponding posture of male goldeneyes, in which the head and neck are stretched forward in the water as the opponent is faced.

Male pair-forming displays include a "neck-erect-forwards" posture, perhaps derived from the alarm posture but differing in that the neck is greatly thickened. A "false-drinking" is frequently performed by males; this display as well as "water-twitching" and preening movements are probably more closely associated with copulatory behavior than with pair-forming behavior. Other movements also occur, such as stretching, bathing, and wing-flapping, but it is uncertain whether these represent ritualized displays or are simply functional body maintenance activities.

Some persons, such as Rawls (1949), have reported hearing males utter vocalizations while courting, and Alex Linski (pers. comm.) also heard a captive male calling during courtship displays. However, Myres (1959a) thought that males were silent during pair-forming display, and other observers have heard no male vocalizations in any context (Brown and Fredrickson 1997). Yet, adult males have complex tracheal structures that differ between the sexes (Johnsgard, 1961) and vary geographically, which suggests that male vocalizations are almost certainly present in some social interactions.

Female displays include a chin-lifting movement, and a very thin whistled note that strongly resembles the inciting behavior of scaup is associated with this action. Chin-lifting is directed toward particular males, suggesting a sexual function. Male-to-male threats or actual attacks are typical during these activities, thus female chin-lifting would seem to represent a functional inciting display (Myres, 1959a).

Copulatory behavior. Myres (1959b) has described the copulatory behavior of this scoter, based on five observed instances. The female apparently assumes the prone position only immediately prior to the male's mounting. Before copulation, false-drinking was performed by the male alone, or by both sexes mutually. Additionally, the male performed preening-behind-the-wing, preening of the dorsal region, or preening along the flanks, either on the side toward the female or the opposite side. Preening movements were seen more frequently than "water-twitching" movements of the bill, but whenever water-twitching was seen, it was always followed by preening. After mounting, the male might perform a flicking movement of the wings.

Postcopulatory behavior is evidently quite simple. Myres noted that on two occasions a partial rotation of the two birds occurred before the male released his grip, and in no case was a specific posturing of the male observed at this time. P. W. Brown (in Brown and Robinson, 1997) offered confirming observations, noting that the male often flicked his wings on dismounting and held on to the female's nape for a few seconds, which could produce rotary movements in the two birds, but no specific postcopulatory displays were observed.

Nesting and brooding behavior. In spite of their late arrival at the breeding grounds, there does not appear to be a rapid transition to nesting behavior. Vermeer (1969) noted in Alberta during 1965 that 36 days elapsed between spring arrival and the laying of the first egg. Rawls (1949) likewise found in the Delta, Manitoba, region, that scoters usually arrived during the first half of May, but nests were evidently not started until the first half of June. This late nesting initiation, and the species' notably long incubation and fledging periods of the species, would seemingly place a restriction on the birds' northward breeding limits.

During the egg-laying period, the male remains with the female, except during the times that she is on the nest. Females do not cover their eggs with down or other materials during the egg-laying phase, and the highest observed nest mortality rates occurred at this time (Hildén, 1964). During the incubation period small groups of males became progressively more frequent, and they usually left the breeding area by the time the young had hatched. In a few isolated instances males have been seen participating in brood care and in protecting the young from gull attacks (Bauer and Glutz, 1969).

After hatching, females take their broods to suitable habitats, which in coastal environments consist of shallow and narrow water areas well sheltered from rough seas. However, the young often do not stay long in the sheltered bays; perhaps as their food requirements change to larger animals they move to areas where kelp beds provide ample foraging habitat for mollusks and crustaceans.

Where populations are dense and suitable brood habitats are limited, massive merging of broods often occurs, with aggregations of 100 or more young not uncommon. The loose female-young bond and the tendency of females to leave their young for prolonged periods also facilitate such brood mergers. However, reductions of such large broods also commonly occur, sometimes apparently being caused simply by strange females swimming nearby and stimulating a "following" response on the part of some of the ducklings (Hildén, 1964).

Postbreeding behavior. Rawls (1949) observed that by mid-July in the Delta, Manitoba, region the males began to band together on Lake Manitoba. These flocks contained from 8 to 20 birds, whereas by mid-August many individual and apparently flightless males were seen. The first flightless females were seen at the end of August, although some hens were still leading broods at that time. The absence of immatures in the area at that time suggested to Rawls that they might have molted earlier and had already begun their fall migration. However, Mackay (1891) noted that adults of all three species of scoters arrive in fall along the coast of New England several weeks prior to the arrival of young birds.

Postbreeding molting areas have not been well studied, but Herter et al. (1989) observed about 66,000 scoters flying past Alaska's Cape Pierce from late June to late July, 77 percent of which were white-winged scoters. These observations suggest that a major molting area might exist along the coastlines of western Alaska or eastern Russia. However, some females in Saskatchewan molt on their breeding areas (Brown and Fredrickson, 1989).

Bufflehead
Bucephala albeola (Linnaeus) 1758

Other vernacular names. Butterball, buffalo-head

Range. Breeds from southern Alaska and northern Mackenzie District through the forested portions of northern Canada east to James Bay and south into the western United States to northern California, Montana, and North Dakota. Winters along the Pacific coast from the Aleutian Islands south to central Mexico, along the Gulf and Atlantic coasts from Texas east to Florida and Atlantic seaboard north to southern Canada, and in the continental interior wherever open water occurs.

Subspecies. None recognized.

Measurements. *Folded wing:* Delacour (1959): Males 163–180 mm, females 150–163 mm.
Culmen: Delacour (1959): Males 25–29 mm, females 23–26 mm.

Weights (mass). Nelson and Martin (1953): 23 males ave. 1.0 lb. (453 g); 26 females ave. 0.7 pound (317 g); max. of both sexes 1.3 lb. (589 g). Yocom (1970): 62 males (August) ave. 14.34 ounces (406.5 g); 10 females ave. 10.4 ounces (294.8 g). Erskine (1972): Males average about 450 g and females about 330 g. Over the year, buffleheads are heaviest during fall migration and lightest during the winter.

Identification

In the hand. The smallest of all the North American diving ducks, this is the only species that has a lobed hind toe, an adult folded wing measurement of 180 mm or less, and a tail of less than 80 mm. The very short (culmen length 23–29 mm), narrow bill is also distinctive, and there is always some white present behind the eye.

In the field. In spite of their small size, male buffleheads in nuptial plumage can be seen for great distances; their predominantly white plumage sets them apart from all other small ducks except the extremely rare smew. The disproportionately large head with its white crest is also apparent, especially when the crest is maximally spread. The tiny female is much less conspicuous and is usually seen only after sighting the male, when its small size and white teardrop or oval marking behind the eye provide identifying field marks. In flight, buffleheads are more agile than most other diving ducks, and their small wings, which are dusky below, beat rapidly and flash the white speculum and upper wing-covert coloration. They are likely to be confused only with hooded mergansers when in flight, but the shorter, rounded head as well as the shorter bill set them apart from this species quite easily. Both sexes are relatively silent, even during courtship display.

132

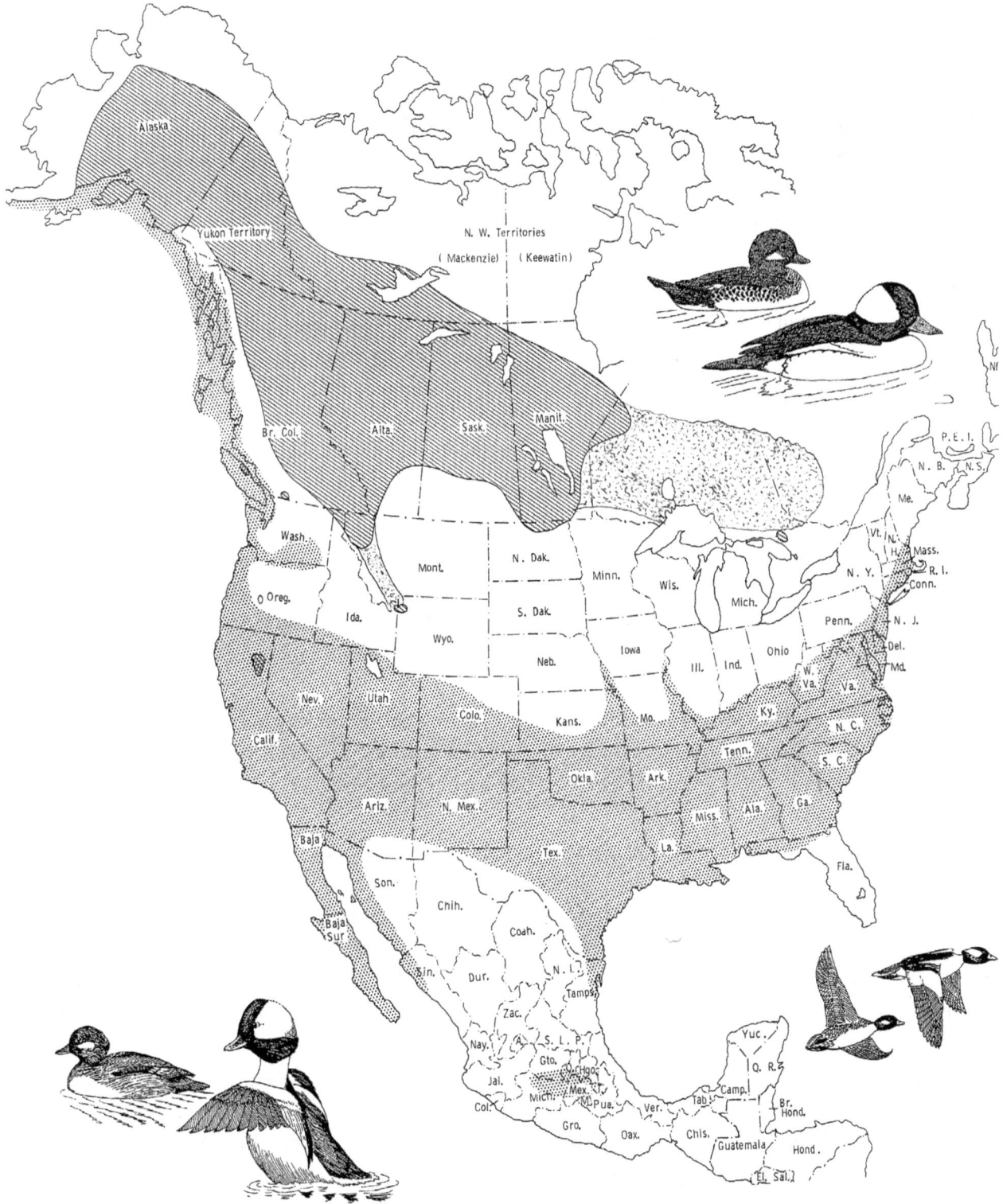

The North American breeding (hatched) and wintering (shaded) ranges of the bufflehead. Regions of newly acquired or previously unrecognized breeding are shown by stippling.

Age and Sex Criteria

Sex determination. During their first year, male buffleheads are difficult to distinguish externally from females, but by late winter the white head markings are larger than those of a female and the male's head is generally darker. After the end of the first year of life, the presence of white in the middle coverts will separate males from females, even during the eclipse plumage. Possibly immature males can be distinguished from females during their first fall and winter by their *flattened* wing measurement of at least 160 mm (Carney, 1964).

Age determination. By their second fall, males will have acquired white feathers in the middle coverts, while first-year males are still black or brownish black in this area. Adult females have tertials which are long and have slightly drooping to rounded tips, whereas immature birds have shorter, straighter ones that are usually frayed and pointed (Carney, 1964). Henny, Carter, and Carter (1981) reviewed additional sex and age criteria for buffleheads.

Distribution and Habitat

Breeding distribution and habitat. The breeding distribution of this North American hole-nesting species is associated with temperate to subarctic forests. In Alaska it is apparently rather widespread through the interior, with its greatest abundance in the upper Kuskokwim Valley, the Yukon flats, and the Porcupine River. Breeding also extends south to the Gulf of Alaska and west perhaps as far as the Bering Sea coast (Gabrielson and Lincoln, 1959).

In Canada the bufflehead breeds in the southern Yukon, western Mackenzie District, east of the Cascades in British Columbia, and across the forested portions of Alberta, Saskatchewan, and Manitoba to northwestern Ontario, where it is local and sparse. Its easternmost breeding would seem to be in Quebec (Gauthier, 1993).

South of Canada there are only a few states that support breeding buffleheads. Although regular breeding in Washington would seem probable, there apparently is so far only one definite record, for Hanson Lake (Larrison and Sonnenburg, 1968). In Oregon buffleheads are local and uncommon breeders in the central and southern Cascades, nesting at high-altitude lakes in Linn, Deschutes, Klamath, and Douglas Counties (Gilligan et al., 1994). Buffleheads have also bred at Eagle and Feather Lakes, Lassen County, California (Grinnell and Miller, 1944) and are local in northern California (Wienmeyer, 1967). In the Rocky Mountains, buffleheads nest at least as far south as northern Montana (Marks, Hendricks, and Casey, 2016), and in Wyoming at least as far south as Grand Teton National Park (Johnsgard, 2014). They also breed locally in northwestern Colorado, where an estimated 50 to 100 pairs nest in Jackson County (Kingery, 1998). In North Dakota buffleheads are regular breeding birds in the Turtle Mountains (Stewart, 1968) and have also been reported breeding in south-central North Dakota (Knutsen and King, 2004). A few breeding records have also been reported for South Dakota, Minnesota, Wisconsin, Iowa, Vermont, Massachusetts, and Maine.

Bufflehead, pair in breeding plumage

During the breeding season their favored habitat consists of ponds and lakes in or near open woodland (Godfrey, 1986). The presence of suitably-sized nest cavities, such as those made by flickers (*Colaptes*), no doubt contributes substantially to the suitability of an area for nesting. An availability of summer insect foods in the form of water boatmen, aquatic beetles and their larvae, and other insects might also be of special importance. Alkaline ponds, sloughs, and small lakes, which are often rich in invertebrate life, are favored over large lakes and high mountain ponds, and trees having suitable nesting cavities should be either surrounded by water or very close to its shore (Munro, 1942). Erskine (1960) reported that eutrophic lakes of moderate depth, having sparse reed beds, generally open shores, and available nest sites such as flicker cavities, are favored for nesting.

Wintering distribution and habitat. In Alaska, buffleheads winter abundantly in the inland bays of southern and southeastern Alaska, westward to the tip of the Alaska Peninsula, and in smaller numbers throughout the Aleutian Islands (Gabrielson and Lincoln, 1959). They are likewise abundant along the coast of British Columbia and inland in the southern parts of that province to the vicinity of Okanagan Lake (Godfrey, 1986), southward through Washington, Oregon, and California, and moderate numbers of birds reach Baja California, Sonora, and northern Sinaloa (Leopold, 1959). In the interior of Mexico and the southern

United States they are present in relatively small numbers. Buffleheads wintering in the Central and Missississippi Flyways have constituted only about 10 percent of the continental population inventoried during recent decades, whereas the Pacific and Atlantic Flyways have supported about a third and a half, respectively.

On the Atlantic coast buffleheads become progressively more common from Florida northward toward the middle and north Atlantic states, with some birds wintering as far north as coastal Maine. In the Chesapeake Bay area they occur on the various types of open estuaries, with brackish estuarine bays providing the optimum habitat. Interior impoundments are also used to some extent, and the birds seem to move farther up small tributaries and inlets than do common goldeneyes (Stewart, 1962).

General Biology

Age at maturity. In captivity, hand-reared buffleheads breed when two years old (Charles Pilling, pers. comm.). Erskine (1961) mentioned three females that were banded as flightless young birds and recaptured on nests two years later. He also has stated (pers. comm.) that wild males regularly breed when two years old.

Pair-bond pattern. Pair-bonds are normally renewed yearly, after a period of social display that begins as early as late January (Munro, 1942). However, Gauthier (1987) reported that two of three marked bufflehead pairs remained together for two years, and one for three years, He also noted that multiple-year retention of mates has been reported in the Barrow's goldeneye, long-tailed duck, harlequin duck, and common eider. Erskine (1961) believed that pair formation might be delayed until the breeding areas are reached. Males leave the breeding area before the young have been hatched (Munro, 1942).

Nest location: Typically, buffleheads nest in tree cavities that have been excavated by woodpeckers, particularly flickers (Erskine, 1960, 1961). Preferred trees are those that are dead and either are situated very close to water, or are standing in water (Munro, 1942). As a reflection of the flicker's preference for excavating in soft woods, nests are most often placed in aspens or, less frequently, rotted Douglas firs. Erskine found that about half of the nest sites used during one year were used again the following year, although at least in some cases different females were involved in the subsequent year. Of 35 females retrapped on nests in later years, 23 were using the same nest site as previously. Brooks (1903) mentioned that cavities in aspen trees used by buffleheads ranged from 5 to 20 feet above the ground, whereas Erskine (1960) found that most nest cavities were from 1 to 3 meters above the ground or at water level. In a few instances nests have been reported to be as high as 40 feet above ground.

The entrance sizes suitable for buffleheads are remarkably small; Erskine (1960) reported that the modal entrance diameter range for natural bufflehead nests was 5.7 to 7.6 cm. The modal cavity depth range was 25 to 37 cm, and the modal cavity diameter range was 11.4 to 16.5 cm. Artificial nesting boxes devised by Charles Pilling (pers. comm.) for captive buffleheads are similar, with entrances 2.5 inches wide, an internal diameter of about 7 inches, and a cavity depth of 16 inches. These boxes were made by splitting logs and hollowing them with a chain saw, followed by wiring them back together.

Bufflehead, males

A sawdust bed about four inches deep was also provided, and the box was situated with its lower end only a few inches above water.

Clutch size. According to Erskine (1960), the average clutch size for initial nests is 8.6 eggs. The observed range for British Columbia nests was 5 to 16 eggs, but clutches in excess of 12 eggs are probably the result of dump-nesting (Godfrey, 1986). Eggs are laid at varying intervals, but average about 38 hours between eggs. The average clutch size of five possible renestings was 6.8 eggs (Erskine, 1960).

Incubation period. The incubation period is reported as ranging under natural conditions from 29 to 31 days, with a modal period of 30 days (Erskine, 1960).

Fledging period. The fledging period was estimated by Erskine (1960) to be 50 to 55 days for wild birds, which is a fairly long period for such small birds.

Nest and egg losses. Erskine (1960) reported that, in his observations, about 80 percent of 106 nests were successful in hatching one or more young, and that about 92 percent of the eggs in 76 successful nests were hatched. Desertion accounted for most of the nest losses, and infertility or embryonic deaths were responsible for most egg failures. Erskine (1964) also reported nest-site competition among buffleheads, mountain bluebirds, and tree swallows. In one case a bufflehead nest was temporarily used by a mountain bluebird, reclaimed by a female bufflehead, and, finally, after the bufflehead had been caught and banded, was taken over by a tree swallow.

Juvenile mortality. Specific information about juvenile mortality is still unavailable. Brood size counts are doubtful indexes to prefledging mortality rates because of the apparent frequency of brood abandonment and brood mergers (Munro, 1942). Erskine (1960) presented data suggesting that the average size of newly hatched bufflehead broods was 8.0 young, as compared with 4.8 young in broods nearly ready to fledge, a 40 percent reduction. His data indicated that large yearly variations in brood survival might occur, with predation, parasites, and drowning all apparently playing potentially significant roles.

Adult mortality. On the basis of recoveries of banded birds, Erskine (1972) calculated an annual survival rate of 46.6 percent for adults of both sexes combined, whereas the calculated survival rate for immature birds was 27.8 percent. On the basis of recaptures of banded females on nests in subsequent years, an adult female survival rate of 50 percent was indicated. Two instances were found of birds surviving for as long as 9 years after banding.

General Ecology

Food and foraging. The 1972 study by Erskine is the most complete food and foraging analysis available for this species. Samples obtained during spring from birds on fresh waters indicated that the larvae of midges and mayflies are commonly eaten, and insects collectively represented two-thirds of the food by volume. In summer, insects constituted more than 70 percent of the total food consumed by adults and virtually all of the food found in 35 downy young. The aquatic larvae (naiads) of dragonflies and damselflies, and aquatic beetles, are important foods for downies, whereas water boatmen and the naiads of dragonflies and damselflies were most prevalent in adult samples.

Only in autumn and winter did plant materials attain a significant proportion of the food samples taken from birds on fresh to moderately brackish waters, when they constituted about 30 percent of total food contents. These plant materials were predominantly made up of the seeds of pondweeds (*Potamogeton* and *Najas*) and bulrushes (*Scirpus*). Insect materials still made up the bulk of the diet in autumn samples, including the sources already mentioned as well as the larvae of mayflies and caddisflies. Winter samples were taken on both freshwater and saltwater areas, and some differences in foods taken in these two habitats were found.

On freshwater wintering habitats, insect foods made up about one-third of the sample volumes, while mollusks increased accordingly and constituted about one fourth of the total. Both gastropod snails and small bivalves such as *Sphaerium* apparently are important winter foods, at least in some areas. On saltwater habitats, insects are largely replaced by crustaceans as major food sources, although mollusks also remain important. Small crustaceans, including decapods, amphipods, and isopods, are apparently the favored winter foods and are supplemented by bivalve and univalve mollusks. Wienmeyer (1967) examined 102 bufflehead

Bufflehead, male and female

stomachs from California and noted that bivalve mollusks, fish eggs, fish, snails, insects, seeds, and vegetative matter all might be locally important foods, depending on the area in which the birds had been foraging.

In general, buffleheads seem to prefer foraging in water from 4 to 15 feet in depth and, like the goldeneyes, tend to inhabit larger and more open bodies of water (Cottam, 1939; Phillips, 1925). Munro (1942) noted that buffleheads usually forage in small groups, and they generally remain submerged from 15 to 35 seconds during foraging dives. Cronan (1957) noted that seven dives averaged 24.1 seconds, as compared with 30.3 seconds for three dives by common goldeneyes, and Erskine (1972) reported similar diving times for buffleheads.

In a ten-study comparison of bufflehead foods (six freshwater, four saltwater), Gauthier (1993) compared gizzard content samples. Crustaceans and plant materials were found in all of ten of the studies, mollusks were in nine, insects were in seven, and fishes were in six.

In saltwater sites, decapod crustaceans (especially shrimp) were of greatest importance, followed by isopods, amphipods, and other crustaceans. Mollusks were of secondary importance and almost exclusively comprised gastropods (especially snails) and pelecypods. Fish (mostly sculpins and ratfish) and plant seeds (largely pondweeds and bulrushes) were present in most studies.

In six freshwater sites, insects predominated, but crustaceans (especially amphipods) occurred in all six, as did mollusks, primarily gastropods. The insects were diverse and present in all seasons, but Odonata naiads, chironomid (midge) larvae, and water boatmen were notably well represented, as were the aquatic larvae of caddisflies and mayflies.

Sociality, densities, territoriality. Erskine (1960) reviewed the concept of territoriality as it might be applied to buffleheads and concluded that a defense of the female seemed a more appropriate description than the defense of a territory. The nest location was often well removed from the "territory" occupied by the pair, which is usually on ponds or larger bodies of water, the smallest of which was found to be an acre.

Nests are usually well spread out, although three cases of trees having two simultaneously occupied nests were found. Excluding such cases, the minimum distances between nests was found to be approximately 100 meters. In 1958, Watson Lake was found by Erskine to support 18 bufflehead nests, while in 1959 a total of 19 nests or broods were determined to be present. The approximate surface area of this lake was about 450 acres; thus a density of about 1 breeding pair per 25 acres of water was present during these years. Not all of the lake's shoreline was used by buffleheads, and counts of both females and males on the lake never tallied with the known nesting population.

Interspecific relationships. The small size and insect-eating tendencies of this species rather effectively remove it from competition with other diving ducks for food or nesting sites. There is, however, a dependence on woodpeckers for providing adequate nesting sites, and varying degrees of competition for these sites exist with various hole-nesting bird species. Besides the common starling, mountain bluebird, and tree swallow, the larger American kestrel, northern hawk-owl, and northern saw-whet owl might also compete locally with buffleheads for nest sites. The Barrow's goldeneye nests in the same areas as the bufflehead but selects cavities with larger entrances (usually over 7.6 cm) and wider cavity diameters (usually over 16.5 cm), according to Erskine (1960).

General activity patterns and movements. Like the other diving ducks, buffleheads are daytime foragers. Probably there are only quite limited daily movements associated with such foraging activities, although specific data are lacking. Erskine (1961) reported a rather surprising degree of homing to previously used nesting areas on the part of females, and to previously used wintering areas on the part of both sexes. Of 12 determinations of nesting location changes in renests or nestings in subsequent years, the average distance between the nesting sites was less than 800 meters.

Similarly, among buffleheads banded on wintering areas in Oregon, New York, and Maryland and shot in subsequent winters, 45 were recovered less than 15 kilometers away, 9 were recovered at distances of 15 to 50 kilometers, and 10 from 50 to 80 kilometers away.

Bufflehead, male and female

Social and Sexual Behavior

Flocking behavior. Like the goldeneyes, buffleheads are not highly social, and while on migration, as well as on wintering areas, they tend to remain in quite small groups. The males are surprisingly pugnacious toward one another, and this general level of aggressiveness probably accounts for the rather small flock sizes typical of buffleheads.

Pair-forming behavior. Pair-forming displays have been described by Myres (1959a) and Johnsgard (1965). The male displays associated with pair formation are obviously derived from attack and escape species. Male buffleheads also assume an aggressive "head-forward" posture that has an identical counterpart in the goldeneye species. Likewise, an "oblique-pumping" movement of the head is very frequent and is the comparable display to bowsprit-pumping and rotary-pumping in the goldeneyes. Males often chase one another while performing this display, or it might be directed toward females. At times it is interrupted by a sudden lifting of the folded wings, retraction of the head toward the back, and a down-tilting of the tail, a possible evolutionary homologue of the goldeneye head-throw-kick display.

When near a female, the male will often erect his bushy crest, making his head seem even larger than normal. While in this posture the male often "leads" a female, which typically follows the male while calling and alternately stretching and retracting her neck. During this display the male might also perform quick, inconspicuous head-turning movements and sometimes suddenly turn his head backward toward the female in an inciting-like movement. Bufflehead pair-forming activities are highly animated, with the jerky and quick movements of the males adding a mechanical or toy-like quality to the proceedings. Often a male will take off, fly a short distance toward a female, and come to a skidding stop near her. This flight display is terminated by a wing-flapping accompanied by a single slapping sound, and as a final stage the male raises his folded wings above the body in the manner described earlier.

Males also sometimes attack one another by submerging and approaching the other bird under water, which often produces amusing results. Female displays consist of the previously mentioned "following" movement, which is functionally equivalent to inciting, and of a "head display," which is comparable to the crest erection of the males and seemingly stimulates them to begin social courtship.

Copulatory behavior. The most complete observations on copulation in this species are those of Myres (1959b). Unlike the goldeneyes, female buffleheads rarely remain in the receptive prone posture for more than a few moments. Myres evidently observed no specific mutual behavior prior to the assumption of this posture. However, prior to mounting, the male performed two main precopulatory displays, a lateral movement of the bill in the water ("water-twitching") and a preening movement of the dorsal region ("preendorsally"). The first was more frequent, but both are extremely similar to the normal bill-dipping, dorsal-preening movements observed as nondisplay comfort movements of the species.

In three of eight copulations, Myres observed "wing-flicking" by the male while it was mounted, and after treading was completed the two birds typically rotated a full turn or more before the male released his grip. Postcopulatory behavior by the male was quite varied, consisting of vigorous bathing, shallow diving, or a deep dive under the surface.

Nesting and brooding behavior. Little information on nesting behavior is available, other than the fact that several observers have commented on the female's strong incubation tendency and her frequent reluctance to leave the nest when it is being examined. Possibly the small entrance prevents most avian and mammalian predators from gaining entrance, and thus there is normally little need for rapid escapes.

Erskine (1960) reported that most egg-laying apparently occurred during morning hours, and that after incubation began the highest degree of nest attentiveness was apparently during morning hours, gradually declining through the day. No tendency for a morning break in incubation was found, but birds were often found away from the nest during evening hours.

Male displays of the bufflehead (1–3) and Barrow's goldeneye (4–10), including (1) two phases of oblique-pumping, (2) folded-wing-lifted, (3) bill-pointing, (4) laying-the-neck-on-the-water aggressive posture, (5) crouched posture, (6–7) neck-withdrawing sequence, (8) head-throw-kick, and (9–10) precopulatory drinking and wing-stretching sequence.

Following hatching, the female typically broods the young for 24 to 36 hours before leaving the nest. Departure from the nest usually occurs before noon, and in one case observed by Erskine about 12 minutes elapsed between the exit of the first and last duckling in the brood. The female was extremely active both before and during the nest exodus, but no vocal signals were detected. After the brood has left the nest, brood territories are established and might be occupied for four weeks or longer.

The ducklings are not defended against other bufflehead broods, although female Barrow's goldeneyes sometimes attack and kill young buffleheads. Brood transfers are not uncommon, and sometimes single broods have been seen accompanied by two females (Munro, 1942). The timing of the breakdown of pair-bonds is seemingly still poorly documented. There are no indications that pair bonds are ever still intact at the time of hatching, and Erskine (1972) stated that the males leave their territories as soon as the females begin incubation.

Postbreeding behavior. Buffleheads sometimes move to molting areas well away from their breeding grounds in western Canada. Erskine (1961) reported that 13 females that were banded while molting and 5 banded as juveniles were later taken as molting adults. Fifteen of these were recaptured on the same lake at which they were banded or within 5 kilometers of that point, while the other four were recaptured at points 25 to 65 kilometers from the point of banding. Two of the birds in the latter group were those that had been banded as juveniles. Erskine thus suggested that adult female buffleheads tend to return to the same molting area. In one instance a female was found to have molted 155 kilometers from a later nesting area, indicating that a substantial migration to molting areas sometimes occurs.

Erskine later (1972) reported that flocks of buffleheads, including males in very faded plumage but still able to fly, have been seen on Alberta lakes in areas where no breeding by this species occurred, a further indication of a substantial molt migration. These movements probably normally involve not only adult males but also immature birds, unsuccessful females, and those that have abandoned their broods.

Barrow's Goldeneye
Bucephala islandica (Gmelin) 1789

Other vernacular names. Whistler

Range. Breeds in Iceland, southwestern Greenland, northern Labrador, and from southern Alaska and the Mackenzie District of Northwest Territories southward through the western states and provinces to California and Colorado. Winters primarily along the Pacific coast from Alaska to central California, and on the Atlantic coast from southern Canada to the mid-Atlantic states.

Subspecies. None recognized.

Measurements. *Folded wing:* Delacour (1959): Males 232–248 mm, females 205–224 mm.
   Culmen: Delacour (1959): Males 31–36 mm, females 28–31 mm.

Weights. Nelson and Martin (1953): 3 males ave. 2.4 lb. (1,087 g), max. 2.9 lb. (1,314 g); 7 females ave. 1.6 lb. (725 g), max. 1.9 lb. (861 g). Palmer (1949): 3 males, ave. 1,162 g, max. 1,219 g; 2 females 794–907 g. Yocom (1970): 53 males (August) ave. 2.125 lb. (1,021 g); 14 females ave. 1.31 lb. (595 g).

Identification

In the hand. The presence of white markings on the middle secondaries and their adjoining coverts, yellow feet with a lobed hind toe, and yellowish eyes serve to separate this species from all others except the common goldeneye. Adult male Barrow's goldeneyes are very much like male common goldeneyes but differ in the following features: (1) The head iridescence is glossy purple, and the white cheek marking is crescent shaped; (2) the head has a fairly flat crown, and the nail is distinctly raised above the contour of the gradually tapering bill; and (3) the body is more extensively black, especially on the flanks, which are heavily margined with black, and on the scapulars, which are margined with black on an elongated outer web or both webs, producing a pattern of oval white spots separated by a black background. The Barrow's upper wing surface is also more extensively black, with the exposed bases of the greater secondary coverts black and the marginal, the lesser, and most of the middle coverts also blackish. Only about five secondaries have their exposed webs entirely white, while the more distal ones might be white-tipped. The length of the bill's nail is at least 12 mm in this species, as compared with a maximum of 11 mm in the common goldeneye (Brooks, 1920).

Females are closely similar to female common goldeneyes but might be separable by (1) the somewhat darker brown head, which is relatively flat-crowned in profile; (2) the brighter and more extensively yellow bill during the spring, especially in western populations, where it is usually entirely yellow; (3) the sooty middle and lesser wing-coverts, which are only narrowly tipped with grayish white; and (4) the broader and more pronounced ashy brown breast band. Brooks (1920) reported that the shape of the bill and the length of the nail provide the best criteria, with the common goldeneye having a nail length that never exceeds 10 mm (average 9.4 mm) and the Barrow's goldeneye having a nail length greater than 10 mm (average 10.9 mm).

In the field. A lone female can be separated from common goldeneye females in the field only by experienced observers, but its somewhat darker head with its flatter crown is usually apparent. Any female with a completely yellow bill is most likely to be a Barrow's goldeneye, although Brooks (1920) noted one possible exception to this rule. A male in nuptial plumage appears to be predominantly black in the upperparts to a point below the insertion of the wing, with a row of neatly spaced white spots extending from the midback forward toward the breast, where an extension of black continues down in front of the "shoulder" to the sides of the breast. Its head is more distinctly "flat-topped," with a long nape and a purplish head gloss, and there is a crescent-shaped white mark in front of the eye.

In flight, females of the two goldeneyes appear almost identical (the yellow bill is often quite apparent in the Barrow's during spring), but the white marking on the upper wing surface of the male is interrupted by a black line on the greater secondary coverts. In both goldeneyes a strong whistling noise is produced by the wings during flight, which is the basis for the collective vernacular name "whistler."

In contrast to the common goldeneye, male Barrow's goldeneyes utter no loud whistled notes during courtship; the commonest male sounds are clicking noises and soft grunting notes. Head-pumping movements of the female Barrow's are of a rotary rather than elliptical form, and lateral head-turning or inciting movements are much more frequent in the Barrow's goldeneye. Tobish (1986) provided additional information on distinguishing the Barrow's from common goldeneyes in all plumages. Occasional natural hybrids between them have been found (Martin and Di Labio, 1994).

Age and Sex Criteria

Sex determination. Young males can be distinguished from females as early as the first November of life, with the appearance of new inner scapulars that are white with extended black edges; at about the same time white feathers begin to appear between the bill and the eyes. Thereafter, the sexes can be distinguished either by the white back or head markings or, when the bird is in eclipse, by the pure white middle coverts of the male.

Age determination. Adult males can be distinguished from first-year males by their entirely white rather than gray or dusky middle coverts. The middle coverts of adult females are grayish, tipped with white, while those of first-year birds are grayish, with dusky bases. The presence of a largely or entirely yellow bill is indicative of a mature female, but mature females might not show this trait during fall and early winter. Some first-year females have a "more or less" orange bill by late April or early May (Brooks, 1920). First-year females also have the chest band and flanks more fawn than gray, and the neck is not white as in adults but is almost as dark brown as the head. Additionally, the iris is greenish yellow, rather than clear yellow as in adults (Munro, 1939).

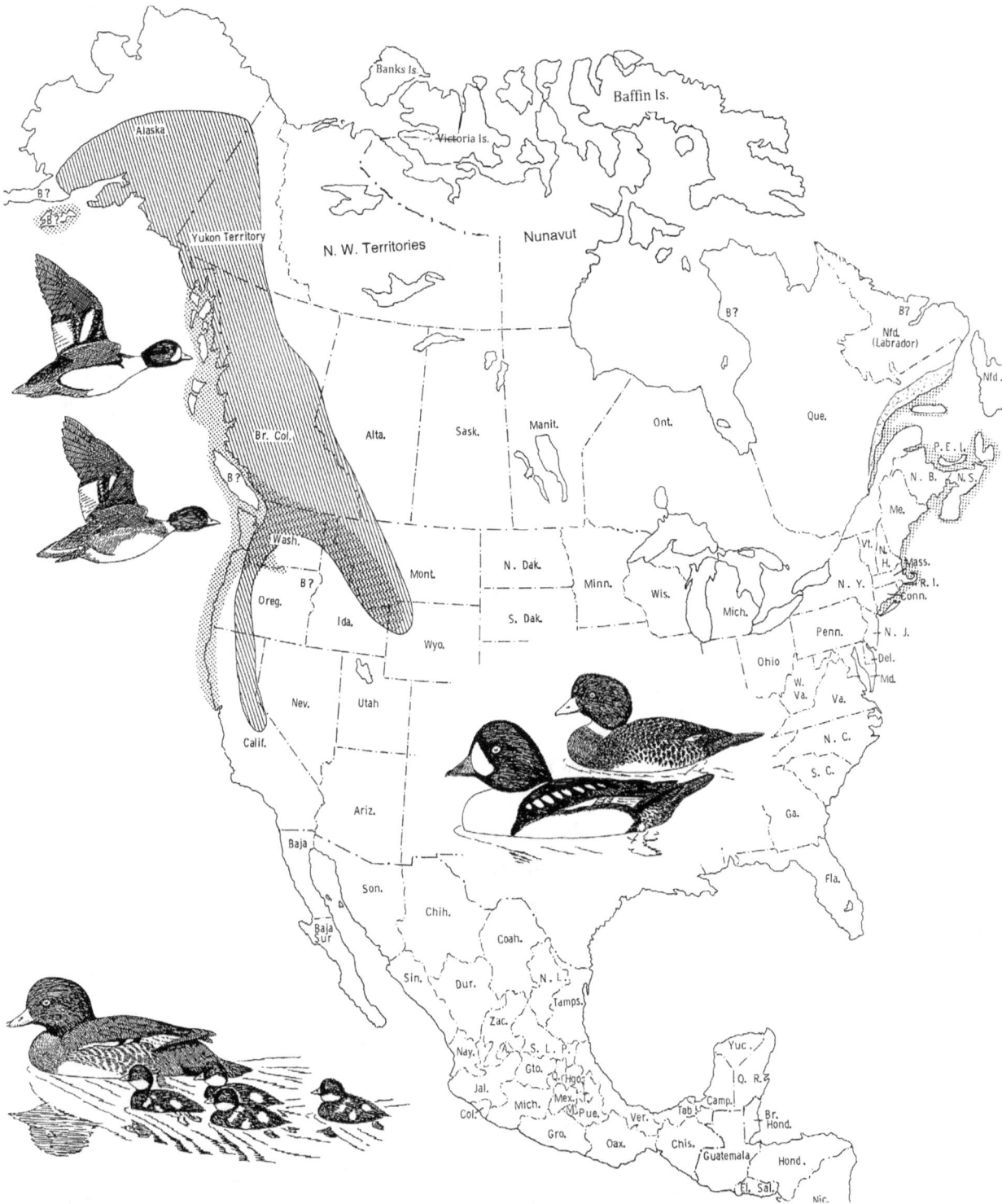

The North American breeding (hatched) and wintering (shaded) ranges of the Barrow's goldeneye. Note that the location of Nunavut is shown on this map only.

Distribution and Habitat

Breeding distribution and habitat. The North American distribution of the Barrow's goldeneye is rather similar to that of the harlequin duck, consisting of a large western population associated with montane rivers and lakes and a much smaller northeastern population in tundra or boreal habitats. In Alaska the breeding distribution of the Barrow's goldeneye is rather poorly understood. It apparently breeds as far west as the base of the Alaska Peninsula (Murie, 1959) around the upper part of the Nushagak River, and extends into the interior northeastwardly through Denali National Park to the Porcupine River. It also breeds on the Kenai Peninsula and in the upper Copper River (Gabrielson and Lincoln, 1959).

In western Canada the species breeds in southern Yukon Territory, southwestern Alberta, and throughout British Columbia. There the densest populations in the relatively dry and sparsely wooded belt between the Okanagan and Cariboo districts, in southern and central parts of British Columbia, where aspen parklands are common. Few breed along the coast (Davidson et al., 2008–2012).

In Washington, breeding has been reported from the Selkirk and the Cascade mountains, the Okanogan Highlands, and the Grand Coulee area in the center of the state (Yocom, 1951; Harris et al., 1954). Breeding occurs sparingly in the mountains of central Oregon on certain lakes (Sparks, Diamond), and possibly also in the Wallowa Mountains (Gabrielson and Jewett, 1940). In California it breeds on various lakes (Butte, Smedberg, Table, etc.) in the mountains southward as far as Yosemite National Park (Grinnell and Miller, 1944). There is also an interior breeding population that extends down the Rocky Mountains of Montana and Idaho southward at least as far as Wyoming's Grand Teton National Park, where it is fairly common on beaver ponds (Johnsgard, 2013). Breeding historically occurred in Colorado during the late 1800s (Bailey and Niedrach, 1967), but it was not until about a century later in 1990 that they were again documented breeding in the Flat Tops Wilderness of northwestern Colorado, at elevations of 10,800 to 11,000 feet (Kingery, 1998).

Savard and Dupuis (1999), Robert et al. (2000), and Robert, Benoit, and Savard (2002) have described this species' little-documented range and breeding status in eastern Canada. Evidently it is most common in the Laurentian Highlands region of southern Quebec, and the total population might be in the range of 3,000 to 4,000 birds. Its core breeding area is in the high plateaus north of the St. Lawrence River from the Saugenay River east. It might also nest in Labrador and northern Quebec, and has been reported during summer at McCormick Island, eastern Hudson Bay (Godfrey, 1986).

The breeding habitat of this species consists of lakes or ponds often in the vicinity of wooded country; where large trees lack natural cavities, rocks might serve for nest sites. In Iceland the species' breeding distribution is largely related to nesting hole availability, but it favors running water over nonflowing water (Bengtson, 1970). Munro (1939) believed that a local abundance of food rather than availability of nest sites determined distribution patterns of this species in British Columbia. Thus, lakes lacking tree-nesting sites but having high populations of amphipods and other foods sometimes support large breeding populations of goldeneyes. These lakes are often rather alkaline and sometimes have relatively little shoreline vegetation.

Barrow's goldeneye, male in breeding plumage

Wintering distribution and habitat. Probably the great majority of the continental population of Barrow's goldeneyes winter along the Pacific coast. From Juneau southward along coastal Alaska the species is common to abundant, and small numbers winter around Kodiak Island as well (Gabrielson and Lincoln, 1959). It also winters abundantly along the coast of British Columbia and more rarely occurs in the interior (Godfrey, 1986). It is common in the Puget Sound region of Washington, where it and the common goldeneye constitute about 9 percent of the wintering diving ducks (Wick and Jeffrey, 1966), and it occurs with decreasing abundance southward along the coasts of Washington, Oregon, and northern California. The birds of the interior Rocky Mountain population do not migrate to salt water but winter near their breeding areas on available open water. Thus, at Red Rock Lakes National Wildlife Refuge, Montana, both Barrow's and common goldeneyes are common in winter and feed on grain put out for trumpeter swans (Banko, 1960).

The wintering population of eastern North America is relatively small, with birds occurring from the north shore of the Gulf of St. Lawrence southward through the Maritime Provinces and the St. Lawrence

Valley, sometimes to the Great Lakes (Godfrey, 1986). Perhaps some of these represent Greenland-bred birds. Surveys in 2009 and 2011 indicated a winter population of about 4,000 to 7,000, as compared with about 23,000 in the central interior of British Columbia (Canadian Wildlife Service Waterfowl Committee, 2013). Elliot (1961) summarized the history of records of this species in New York, and Hasbrouck (1944) provided an earlier summary of its more general distribution during winter in eastern North America.

The wintering habitats used by Barrow's goldeneyes include both fresh water and salt water, with the greatest numbers no doubt occurring in fresh water or brackish habitats. Munro (1939) noted a preference for fresh or brackish rather than highly saline waters, and reported higher numbers on coastal lakes, rivers, and river mouths than on more saline waters. The birds also evidently prefer brackish estuaries and calm waters to open coastlines and heavy surf.

General Biology

Age at maturity. Ferguson (1966) reported that four aviculturists reported initial breeding by captive Barrow's goldeneyes at two years, four reported three years, and one at five years. It seems probable that, at least under ideal conditions, the birds might initially breed in their second year.

Pair-bond pattern. Pairs are typically renewed each year during a prolonged period of social display during winter and spring, with a peak in courting activity during April in British Columbia (Munro, 1939). However, Savard (1985) offered some evidence of long-term pair-bonds in Barrow's goldeneyes, in that several pairs of marked birds were reunited in subsequent years.

Nest location. Bengtson (1970) reported that of a total of 426 nests found in Iceland, 401 were in holes or cavities, 19 were under high shrubs, 5 were under low shrubs, and 1 was under tall *Angelica* or *Archangelica* herbs. The most frequent distance from water to the nests he found was the within-ten-meters category. In British Columbia, cavities in live trees, tree stumps, or tall dead stubs are the usual nest sites (Munro, 1939). Sugden (1963) reported that one of 13 nests he found in the Cariboo Parklands was in a crow nest, while all the others were in holes of Douglas firs or aspens. He suggested that crow nests might be more important at higher elevations, where trees suitable for hole-nesting are less numerous.

Among Barrow's goldeneye tree-cavity nests in British Columbia, Erskine (1960) reported that 16 of 30 nests had entrances 7.6 to 10.0 cm in diameter, 11 of 19 had cavity depths of 25 to 134 cm, and 8 of 14 had cavity diameters of 16.5 to 22.8 cm. In contrast to cavities used by buffleheads, those with top entrances are often used, and such entrances might be preferred to lateral ones.

Clutch size. Godfrey (1966) reported an average clutch size of 9 eggs, with a range of 4 to 15, among nests in British Columbia and Alberta. Bengtson (1971) indicated an average of 10.37 eggs in 293 first clutches in Iceland. He found (1972) an average of 7.5 eggs in 39 renestings.

Barrow's goldeneye, female and brood

Incubation period. Under natural conditions incubation lasts an average of 32.5 days, with a reported range of 30 to 34 days (Godfrey, 1986). A 30-day period has been reported for captive birds, presumably under artificial incubation (Delacour, 1959; Johnstone, 1970).

Nest and egg losses. Slightly over half of 196 Icelandic nests failed during the egg-laying period, with desertion the major cause. Among 246 normal-sized clutches observed over a ten-year period, the estimated hatching success was 75 percent (Bengtson, 1972).

Juvenile mortality. Munro (1939) reported that his studies on one lake in 1936 indicated a reduction (from 9 to 6) in brood sizes of 33 percent among 9 broods during the first month after hatching, whereas in 1937 14 broods had a reduction in numbers of 66.5 percent within about two months after hatching. He believed that crows might account for some duckling losses, and that great horned owls also might contribute to brood losses. Exposure to rough water on some of the larger lakes was also considered a probable duckling mortality factor. Bengtson (1972) observed that female goldeneyes often attacked and killed strange young of their own species, as well as those of other species, and were so aggressive toward others that they often abandoned their own broods.

Eadie et al. (2000) estimated the average breeding lifespan for 81 marked females was 2.16 years in British Columbia, and the average lifetime reproductive success was 3.1 fledged young. However, 69.1 percent of the marked females experienced only 1 to 2 years of breeding. Just 25.5 percent bred 3 to 4 years, and 5.5 percent survived more than four years of breeding. Those females with the greatest number of breeding years were the most successful in reproducing, with 13.6 percent of the females being responsible for the production of 52.67 percent of 251 total ducklings hatched.

Adult mortality. Eadie et al. (2000) calculated that the annual survival rate of 110 breeding females in British Columbia was 66.2 percent, while another estimate based on repeat sightings of 533 marked females was 66.8 percent (Savard and Eadie, 1989). Delacour (1959) mentioned two males surviving at least 16 years in captivity.

General Ecology

Food and foraging. Cottam (1939) analyzed the foods taken by 71 adults, most of which were from British Columbia, over 11 months of the year. Among these foods, insects constituted 36 percent by volume, mollusks 19 percent, crustaceans 18 percent, other animal foods 4 percent. Plant materials totaled 22 percent and appeared to be largely pondweeds (*Potamogeton*) and wild celery (*Vallisneria*). The insect category included large quantities of dragonfly and damselfly naiads, caddisfly larvae, midge larvae, and various other aquatic insects. The major mollusk food was the blue mussel (*Mytilus*) and related mussels, whereas the crustacean category was dominated by amphipods, isopods, and crayfish.

Munro (1939) also reported on the foods of this species, using some of the same data included in Cottam's analysis, as well as additional materials, bringing the total to 116 stomachs. Salmon eggs were found to be an important food source for coastal birds, along with mollusks, crustaceans, and marine algae. Birds taken in interior regions contained a variety of insects, especially caddisfly, damselfly and dragonfly naiads, crustaceans such as amphipods and crayfish, and various plant materials. Several young birds had mainly eaten insects, especially surface-dwelling and terrestrial species. Munro concluded that the winter foods of the Barrow's and common goldeneyes are substantially the same, under the same conditions of time and space.

Sociality, densities, territoriality. Bengtson (1972) has provided evidence that this species was the only one among the ten duck species he studied that exhibited density-dependent relationships in hatching success and duckling mortality. He reported that breeding densities varied from 30 pairs per square kilometer in "scattered" concentrations up to 100 to 600 pairs per square kilometer in "very dense" concentrations. He also observed that this species might be unique in its defense of a territory prior to and during egg-laying, and in females defending brood territories after their clutches had been hatched.

Barrow's goldeneye, pair in flight

Evidently the strong aggressive tendencies among female goldeneyes result in high rates of nest desertion where breeding concentrations are high, and similarly tend to reduce brood survival as a result of attacks on broods sharing the same brooding areas. Dense goldeneye concentrations also tended to reduce brood survival of other duck species.

Interspecific relationships. It seems possible that local competition for food between the Barrow's and common goldeneyes might occur on wintering areas, since there appear to be no significant differences in foods consumed. There also appear to be no major habitat differences that tend to keep the two species separated on wintering areas, although Munro (1939) noted that common goldeneyes are true sea ducks during winter, frequenting the most saline waters, whereas Barrow's goldeneyes favor fresh or brackish waters.

The two species' breeding areas are for the most part isolated from one another, but they overlap in British Columbia, and both species lay their eggs of the other indiscriminately (Eadie, 1989). Phillips (1925) mentioned that in Iceland the eggs of the red-breasted merganser are often found among those of the Barrow's goldeneye. Other species whose eggs have been found among nests of Barrow's goldeneyes include the bufflehead, hooded merganser, and wood duck (Palmer 1976).

The parasitic jaeger is the most important predator in Iceland, but since the nests of goldeneyes are normally well hidden, predators such as jaegers and gulls would no doubt primarily affect the ducklings rather than eggs.

General activity patterns and movements. Like the other sea ducks, Barrow's goldeneyes are daytime foragers, but little specific information on daily activity patterns or movements is available.

Social and Sexual Behavior

Flocking behavior. On the wintering grounds and during spring migration there is a free association of adults and yearlings of both sexes, producing flocks of moderate to large size. Migrant flocks mentioned by Munro (1929) generally ranged from 10 to 40 birds. Within a month after arrival, the adults have paired and scattered. The yearling males gradually disappear during this time, followed by the adult males as soon as the females begin incubation. Many or all of the yearling females remain on the nesting areas, leaving about the time the adult females do, followed finally by the young of the year. At this time there is again a general association of the total population on coastal waters (Munro, 1929).

Pair-forming behavior. Pair-forming behavior begins on coastal waters in late winter but reaches its peak on the lakes of the interior where migrant birds concentrate. The male displays of this species are quite varied and complex (Myres, 1959a; Johnsgard, 1965), differing both in postures and vocalizations from those of male common goldeneyes. These male differences in behavior and appearance, rather than female differences, probably serve to maintain species isolation and prevent hybridization (which has rarely been reported).

Goldeneyes typically perform social display in small groups of several males and one or two females. The male displays, although highly ritualized, are apparently largely derived from hostile gestures of threat of

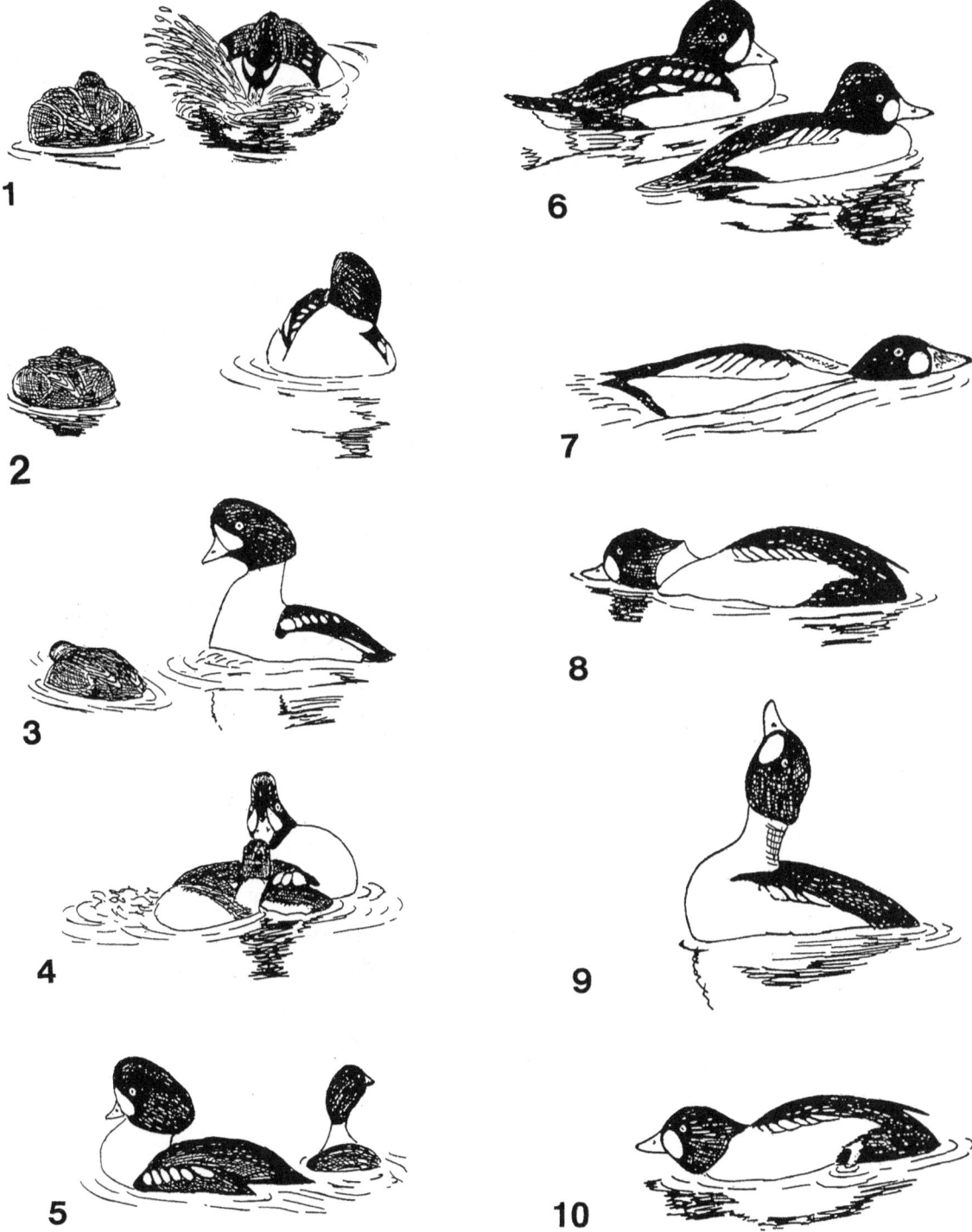

Male displays of the Barrow's goldeneye copulation sequence (1–5), including (1) jabbing, (2) preening-behind-the-wing, (3) steaming to female, (4) rotations, (5) postcopulatory steaming, (6) comparison of Barrow's and common goldeneye male breeding plumages, and (7–10) common goldeneye male (7) laying-the-neck-on-the-water aggressive posture and (8–10) masthead sequence.

Barrow's goldeneye, pair rotary-pumping

attack. The female's primary display, inciting, is also a highly ritualized side-to-side movement of the head, frequently performed as she follows a displaying male. She also performs rotary-pumping movements of the head, similar to those of the male, and a neck-stretching or "head-up" display.

The female displays of the Barrow's and common goldeneyes differ in that the Barrow's females lack or very rarely perform "neck-dipping" but much more frequently perform inciting. Major male display differences include only a single type of head-throwing display by the Barrow's goldeneye (which is always associated with a backward kick), a "crouch" posture that is lacking in the common goldeneye, a "neck-withdrawing" movement that is the usual male response to inciting, and the absence of the common goldeneye's "masthead" and "bowsprit" postures (Myres, 1959a; Johnsgard, 1965).

Copulatory behavior. Myres (1959a, 1959b) and Johnsgard (1965) have described the behavior associated with copulation. Copulation is normally preceded by mutual drinking movements, after which the female becomes prone. The male then begins a long sequence of ritualized comfort movements (wing-and-leg-stretch, dipping the bill in the water and shaking it ["water-twitching"], and bathing), in no apparent order. Finally, the bill-dipping and shaking becomes more vigorous ("jabbing"), are terminated with a single

Barrow's goldeneye, pair rotary-pumping (extreme posturing)

rapid wing-preening movement, and the male then rushes toward the female to mount her. During treading the folded wings are shaken one or more times, and before the male releases the female's nape the two birds typically rotate in the water ("rotations"). The male then swims rapidly away ("steaming") while uttering grunting sounds and making lateral head-turning movements. The copulatory behavior of the two species of goldeneye is much more similar than are their pair-forming behaviors.

Nesting and brooding behavior. Few observations of nesting behavior by this species are available. Munro (1939) noted that the female begins to lose down while laying is in progress, but the amount varies considerably in different nests. Males apparently leave their mates very shortly after incubation begins. Often yearling females become attached to paired adults, and after the nesting female emerges with her brood, the young female might resume her association. At times, the yearling might even take forcible possession of the brood, driving the mother away and capturing at least some of her young. Attacks on ducklings by other females defending their own brood territories are also a serious threat to ducklings, and thus broods in locations where several broods are present might have higher mortality rates than those in which a single brood is being tended (Savard, Smith, and Smith, 1993).

Young broods usually but not always remain relatively intact, often closely following the parent bird in a clustered flank-to-flank group but more commonly following head-to-tail. Carrying of the young on the back while swimming has apparently not been described in this species, although it has been observed in the common goldeneye. Females typically abandon their brood when they are well grown but still unfledged. The females might then fly some distance and join other females prior to molting, or they might molt on their breeding lakes.

Postbreeding behavior. According to Munro (1939), there is an early, massive movement of males from the breeding grounds almost as soon as the females begin incubation. The distance and direction of this molt migration is unknown in British Columbia, but the males presumably move toward the coast. Elsewhere in western Canada, the Old Crow Flats in northern Yukon Territory is a known molting area, as is Ohtig Lake in northeastern Alaska, and Yellowstone National Park in Wyoming (Baldassarre, 2014). In eastern Canada, 12 birds that wintered on the St. Lawrence River estuary migrated to sites in Hudson Bay, Ungava Bay, Baffin Island, Labrador, and to an inland site near the Quebec-Labrador border (Robert, Benoit, and Savard, 2002). In Iceland, however, the adult males molt near their breeding areas (Phillips, 1925).

Common Goldeneye
Bucephala clangula (Linnaeus) 1758

Other vernacular names. Golden-eye duck, whistler

Range. Breeds in Iceland, northern Europe, and Asia from Norway to Kamchatka, and in North America from Alaska to southern Labrador and Newfoundland, and southward through the forested portions of the northern and northeastern United States. Winters in North America from the southern Alaska coast south through the western states to California, the interior states wherever open water is present, and the Atlantic coast from Florida to Newfoundland.

North American subspecies. *B. c. americana* (Bonaparte). American Common Goldeneye. Breeds in North America as indicated above.

Measurements. *Folded wing:* Palmer (1957): Males 230–245 mm, ave. (12) 239 mm; females 210–223 mm, av. (12) 218 mm.
Culmen: Palmer (1957): Males 32–33 mm, ave. (12) 3.3 mm; females 30–33 mm, ave. (12) 31.7 mm.

Weights (mass). Nelson and Martin (1953): 58 *B. c. americana* males ave. 2.2 lb. (997 g), max. 3.1 lb. (1,406 g); 54 females ave. 1.8 lb. (815 g), max. 2.5 lb. (1,133 g). Schiøler (1926): *B. c. clangula:* 6 adult males (December–January) ave. 2.55 lb. (1,158.5 g); 9 first-year males ave. 2.29 lb. (1,037 g); 5 adult females ave. 1.76 lb. (799 g); 5 first-year females ave. 1.63 lb. (747 g).

Identification

In the hand. Males in nuptial plumage or those in their first spring of life have a characteristic oval white mark between the yellowish eye and bill. Mature males, even when in eclipse plumage, are the only North American ducks having the combination of a folded wing length of at least 215 mm and an uninterrupted white wing patch extending from the middle secondaries forward over the adjoining greater, middle, and lesser coverts. Females can be distinguished from all other species except the Barrow's goldeneye by their lobed hind toe, a folded wing length of 190 mm or more, white on the middle secondaries, and their greater coverts (duskily tipped), and at least the adjoining middle coverts more grayish or whitish than are the tertials or lesser coverts.

See the Barrow's goldeneye account for other characteristics that will serve to separate females of these two species; Carney (1983) provided species, age, and sex identification of both goldeneyes using wing characteristics. A few natural hybrids between these species have been reported (Martin and Di Labio, 1994).

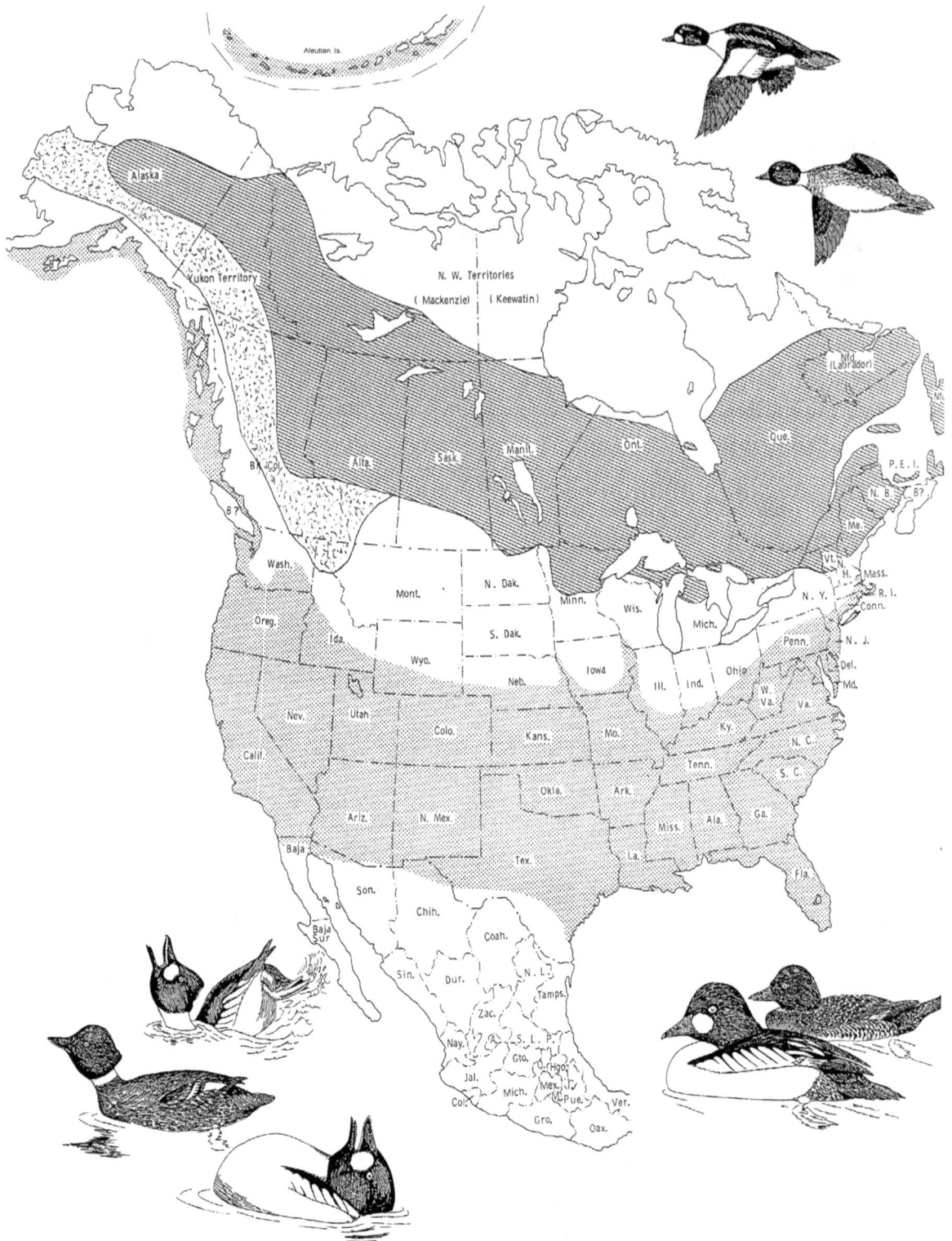

The North American breeding (hatched) and wintering (shaded) ranges of the common goldeneye. Regions of newly acquired or previously unrecognized breeding are shown by stippling.

In the field. Along with the larger and more streamlined common merganser, goldeneyes are the only large diving ducks that appear to be mostly white-bodied, with blackish backs and heads. The oval white mark on the male can be seen for considerable distances, and even if this mark is not visible, the common goldeneye male differs visually from the Barrow's goldeneye male in several other ways. The former's head is more triangular in shape, with the top almost pointed rather than flattened, the nape is not extended into a long crest, and a greenish gloss is apparent in good light. The upper half of the body appears predominantly white, with parallel black lines extending diagonally backward above the folded wing, and no black is evident on the flanks.

In flight, both species of goldeneye exhibit dusky under wing-coverts, but the common goldeneye exhibits a relatively continuous white upper wing patch, at least in males. Females of both species are relatively silent, but the wings of males produce a strong whistling noise during flight, which is less evident in females. The male's calls are varied, but the loudest and most conspicuous is a shrill *zeee-at* note that is associated with aquatic head-throw displays.

Age and Sex Criteria

Sex determination. If the middle secondary coverts are white, the bird is an adult male, and additionally first-year males have less gray on the breast than do females, have a darker back, and a darker head, which might show a white loral spot by late winter (Bent, 1925). First-year males might also exhibit longitudinal white stripes or white edging on some of their scapulars, and if the *flattened* wing (measured from notch in bend of wing to tip of longest primary) is longer than 215 mm the bird is a male, whereas if it is less than 210 mm it is a female (Carney, 1964).

Age determination. Adult males have scapulars with white center stripes and dark edges and have middle coverts that are entirely white, rather than partly white and partly gray as in immatures. The greater secondary coverts of adult females are heavily tipped with black and their greater tertial coverts are rounded and unfrayed over the tertials, while in immature females they are usually both frayed and faded (Carney, 1964). The middle and lesser coverts in immature females are also apparently darker than in adult females. A yellow bill tip is indicative of sexual maturity in females, but this trait might not be apparent during fall and early winter. As in the Barrow's goldeneye, first-year females have greenish yellow rather than clear yellow iris coloration, brownish rather than white necks, and fawn rather than gray flanks and chest bands (Munro, 1939).

Distribution and Habitat

Breeding distribution and habitat. Unlike the Barrow's goldeneye, this species has a breeding distribution extending over much of the cold temperate portions of the Northern Hemisphere. In North America its breeding range is nearly transcontinental, generally following the distribution of boreal coniferous

Common goldeneye, pair in breeding plumage

forest. In Alaska the common goldeneye probably breeds over a wide area of the interior, although confusion over the two goldeneye species makes the distribution of each rather uncertain. Gabrielson and Lincoln (1959) believed that breeding is mostly confined to the Yukon and Kuskokwim River Valleys, excepting the lower portions of these rivers. They are apparently most numerous from Tanana eastward toward the Canadian boundary.

Broods attributed to this species have also been seen on.Kodiak Island, but there seems a greater likelihood that these were of Barrow's goldeneyes. It is also questionable that a brood from extreme southeastern Alaska (Chickamin River) represented this species. Both species occur on the upper Porcupine River in the vicinity of Old Crow, but their relative breeding abundance is unknown (Irving, 1960). Campbell (1969) suggested that the common goldeneye might breed as far north as the central Brooks Range.

In northern Canada this species probably breeds over much of the Yukon north to treeline, over the forested portions of the Northwest Territories, and in the forested areas of all the provinces, with breeding questionable along coastal British Columbia (where Barrow's goldeneyes are very common), and in Nova Scotia except for Cape Breton Island (Godfrey, 1986).

Munro (1939) questioned its widespread occurrence as a nesting species in British Columbia, and aerial surveys are unable to distinguish the two species, so the status of the common goldeneye in British Columbia was long uncertain. However, they are now known to breed throughout British Columbia, except for the Haida Gwaii, but are mostly found in the southern and eastern portions of that province, in the Cariboo and Chilcotin districts (Campbell et al., 1990; Davidson et al., 2008–2012). A recent estimated total provincial population estimate was about 22,000 birds, with about half of the total in the

Common goldeneye, male and female

boreal forest and half in the interior (Baldassarre, 2014). Federal breeding waterfowl surveys in 2010 over western and central North America reported almost 700,000 goldeneyes, a third of which were found in Northwest Territories, British Columbia, and northern Alberta, and another fourth were in northern Manitoba (Baldassarre, 2014).

In central and eastern Canada the breeding range includes much of Manitoba, virtually all of Ontario except the southeastern portion and Hudson Bay coastline, the forested portions of Quebec, and comparable portions of Labrador, Newfoundland, and all the Maritimes (Erskine, 1992). These eastern populations are almost as large as those in the western provinces; breeding population counts in Quebec and Labrador made by waterfowl biologists in the early 1950s indicated that goldeneyes constituted over 20 percent of the nesting waterfowl, surpassed in abundance only by mergansers and black ducks. Federal breeding surveys over the eastern portions of North America from 1990 to 2010 averaged over 416,000 goldeneyes, nearly all of which were common goldeneyes, since only a few thousand Barrow's goldeneyes are thought to breed in the east, and 98.5 percent of the total were found in Canada, primarily Quebec (Baldassarre, 2014).

South of Canada, the common goldeneye has a restricted breeding distribution. They breed in northeastern and northwestern Montana (Marks, Hendricks, and Casey, 2016), the Turtle Mountains area of northern North Dakota (Stewart, 1968), over much of northern Minnesota (Lee et al., 1964a), northern Michigan (Brewer, McPeek, and Adams, 1991), northern New York (Foley, 1960), Maine (Gibbs, 1961), and the northern portions of Vermont and New Hampshire (Bent, 1925).

The preferred breeding habitat of common goldeneyes was described by Carter (1958) as wetlands having marshy shores with adjacent stands of old hardwoods to provide nesting sites. To a much greater degree than the Barrow's goldeneye, this species is limited in its breeding habitats to water areas with trees having adequate cavities for nesting. The depth of the water and whether it is a river or a slough were judged by Carter to be of no importance. In northern coniferous forest areas, aspens are apparently important for nesting, but farther south a greater variety of trees are utilized (Dementiev and Gladkov, 1967).

Wintering distribution and habitat. In Alaska, common goldeneyes are frequent winter residents of the Aleutian Islands (Murie, 1959), as well as around Kodiak Island, the Alaska Peninsula, and the entire coast of southern and southeastern Alaska. They are especially abundant in the vicinity of Wrangell, and around the northern tip of Admiralty Island (Gabrielson and Lincoln, 1959).

In Canada, they winter on the coast of British Columbia, as well as around Newfoundland and the Maritime Provinces, with smaller numbers occurring through the interior where open water is available. South of Canada, they extend in winter south along the Pacific coast to the Mexican border and beyond, although only relatively few birds move that far south. Winter surveys during the late 1960s indicated that about 29 percent of the wintering goldeneye population (both species) were found in the Pacific Flyway, and many of these would have been Barrow's goldeneyes.

The Atlantic coast apparently represents the primary wintering area; almost half of the continental wintering goldeneyes surveyed in the late 1960s were found in the Atlantic Flyway, and very few of these would have represented Barrow's goldeneyes. Common goldeneyes are common along the Atlantic coastline as far south as Florida. They also occur on the Gulf Coast, but in relatively low numbers. Furthermore, goldeneyes winter in the interior of the United States on rivers, large lakes, and reservoirs as far north as open water can be found.

In the Chesapeake Bay region, wintering common goldeneyes are widely distributed in coastal estuaries, but optimum habitats are apparently brackish estuarine bays, with large numbers also using salt estuarine bays. Fresh estuarine bays are primarily used by fall migrants. Munro (1939) believed that, whereas the Barrow's goldeneye favored fresh or slightly brackish waters in its winter habitats, the common goldeneye was prone to frequent more saline waters.

General Biology

Age at maturity. Aviculturists contacted by Ferguson (1966) reported reproduction by captive common goldeneyes at two years (two cases), three years (four cases), and four years (one case). Most investigators

Common goldeneye, male in breeding plumage

believe that nesting normally occurs among wild birds in their second year of life, and some yearling females might even attempt to nest occasionally (Grenquist, 1963).

Pair-bond pattern. Pair-bonds are renewed each winter and spring, during a period of active social display that lasts several months. In Sweden this begins in December and extends into May but has an average peak of activity in March (Nilsson, 1969). Pair-bonds are broken shortly after the female begins incubation.

Nest location. Carter (1958) stated that all but one of the nests he found (apparently 17 in total) were in trees, and 89 percent were in maple (*Acer* spp.) cavities with a diameter at breast height of at least 8 inches. Sixteen nests averaged 18 feet above the ground or water level. Most cavities had lateral openings, but the height and entrance size seemingly mattered very little.

Prince (1968) reported on 16 goldeneye nesting sites, of which 9 were in silver maples (*Acer saccharinum*), 6 were in elms (*Ulmus americanus*), and one was in a butternut (*Juglans cinerea*). These were in trees with an average diameter at breast height of 67 centimeters and an average height of 7 meters. Ten of the cavities were of the "bucket" type and six were enclosed. Trees used by goldeneyes tended to be in open stands near the edges of fields or marshes, and cavities selected by them tended to vary less in cavity diameter than did those used by wood ducks. This small observed range of cavity diameter (15–26 centimeters) was considered to be possibly important in nest site choice. Siren (1951) recommended nesting boxes with similar internal diameters of 18 to 22 centimeters for goldeneyes.

Clutch size. Lee et al. (1964a) reported that the average clutch size of 39 Minnesota nests was 10.2 eggs. Grenquist (1963) reported an average of 10.3 eggs in 53 clutches. Carter (1958) observed a range of 7 to 12 eggs in 9 New Brunswick nests, with an average of 9.0. The egg-laying interval is about 1.5 days (Bauer and Glutz, 1969). Excluding clutches affected by parasitized nests, Eadie, Mallory, and Lumsden (1995) calculated an average clutch of 7.13 eggs.

Incubation period. Probably 27 to 32 days, with a report of 30 days in one case (Bauer and Glutz, 1969).

Fledging period. Reported by Siren (1952) as 61 to 66 days. Lee et al. (1964a) estimated a 56- to 57-day fledging period.

Nest and egg losses. Prince (1968) reported a nesting success of 60 percent for six "enclosed" nests versus only 16 percent for ten "bucket" nests in New Brunswick. Carter (1958) observed a hatching success of 96.2 percent for 79 eggs, excluding one flooded clutch. Sources of egg losses are not well known, but predation levels of enclosed nests are probably low, with rare losses to pine martens. However, in rainy years, soaking of the nest might cause nest abandonment by a large percentage of the females (Dementiev and Gladkov, 1967).

A study by Grenquist (1963) in Finland provides the best information on prehatching losses. Of 1,554 eggs, 50.6 percent hatched, 40.4 percent were not incubated, 6.2 percent had dead embryos, and 1.9 percent were infertile. This low hatching success was attributed to competition for nest sites and particularly to the disruptive effects caused by females laying for the first time and entering previously occupied nests. These were probably two-year-old females, or perhaps those that had come into reproductive condition later than older birds. In clutches that were incubated to completion, an average of 9.6 ducklings hatched.

In North America, most nest failures in a British Columbia study were caused by desertion during egg-laying, while 33 percent failed during incubating (51.2 percent of 80 failed nests); nest desertion was also the primary cause of nest losses in Ontario (Eadie, Mallory, and Lumsden, 1995).

Increased disturbance at the nest site from other females is apparently a major cause of nest desertion. There is often much competition for nest space with other cavity-nesting ducks, and mixed clutches involving common goldeneyes have been found with the eggs of Barrow's goldeneyes, buffleheads, wood ducks, common mergansers, and (in Europe) smews (Eadie, Mallory, and Lumsden, 1995).

Juvenile mortality. Gibbs (1962) reported that, of an original population of 77 young, only 22 ducklings reached the age of fledging, representing a 71.5 percent prefledging mortality. However, specific mortality factors could not be determined. Carter (1958) estimated a 47 percent mortality loss prior to fledging, based on reductions in average brood sizes that he observed over a six-year period, resulting in an average brood size of 4.8 young in broods approaching flight stage. Predation, disease, and accidents were assumed to be responsible for these losses, but specific mortality factors were not identified.

Reproductive success among common goldeneye females is highly skewed, with the majority of females in one nine-year study (Eadie, Mallory, and Lumsden, 1995) producing no independent young, and more than three-fourths of 243 fledged young being produced by only 23 percent of the breeding population. Associated with this statistic was the low average percentage of young (36.5 percent) surviving to independence, and a small average number of young (1.3) that was raised to independence in a single year per nesting female.

Adult mortality. Nilsson (1971) judged that adult goldeneyes in Sweden have an annual mortality rate of about 37 percent (63 percent survival). Johnson (1967) estimated a 36 percent loss of first-year birds as a result of hunting, but a considerably lower adult mortality caused by hunting was estimated, resulting from the adults using large lakes that offered some protection.

Annual survival estimates of nesting adult females based on subsequent-year recaptures, include 61.2 percent among 85 nesting females in British Columbia, and an estimated total lifespan of 3.09 years, including 1.58 year of breeding. Similarly obtained annual survival estimates from four other locations have ranged from 58 to 74 percent. Maximum known ages for birds banded in Canada are 11 years for males and 12 years for females (Eadie, Mallory, and Lumsden, 1995).

General Ecology

Food and foraging. Cottam (1939) reported on the foods found in 395 common goldeneye stomachs obtained during every month except June and August. Crustaceans (32 percent by volume), insects (28 percent), and mollusks (10 percent) constituted the primary animal foods, while a variety of plant foods totaled 26 percent. Favored foods appear to include crabs, crayfish, and amphipods among the crustaceans; the larvae of caddisflies, water boatmen adults, and the naiads of dragonflies, damselflies, and mayflies among the insect foods; and various bivalve and univalve mollusks. Thirteen juvenile birds had been foraging primarily on insects, of which beetles and immature stages of caddisflies, dragonflies, and damselflies were the most important.

Reporting on birds obtained in Danish waters between October and February, Madsen (1954) found that crustaceans, bivalves, and univalve mollusks were the three most important categories of foods present, with insects playing a very minor role in these marine and brackish-water samples. A similar indication of the relative importance of these food sources was given by Stewart (1962), on the basis of 23 birds obtained in the Chesapeake Bay region. Olney and Mills (1963) found comparable differences between birds collected in freshwater and marine habitats, noting the wide range of foods and habitats that are utilized by this species.

In a nine-study comparison of food volumes found in the esophagi, crops, and gizzards of adult common goldeneyes, crustaceans were present in eight of the nine studies, as were mollusks. Fish (eggs, fry, adults) and plant materials or seeds were both reported in seven studies. Insects and their larvae were surprisingly well represented in both diversity and volume, including Trichoptera larvae (seven studies), Ephemeroptera nymphs (five studies), Coleoptera larvae and adults (five studies), Chironomidae larvae (five studies), Corixidae adults (four studies), and Odonata nymphs (three studies). Many of the same insect groups were present in five studies of duckling foods, including nymphs of Ephemeroptera, Anisoptera, Zygoptera, and Trichoptera, larvae and adults of Coleoptera and Diptera, and adults of Corixidae and Notonectidae (Eadie, Mallory, and Lumsden, 1995).

Cottam (1939) commented on the diversity of foods consumed by this species and mentioned that it seemed able to survive on almost any type of available food. He observed these birds foraging in depths of 4 to 20 feet. Nilsson (1969) noted that goldeneyes preferred foraging in waters less than 1.5 meters deep and in deeper waters only when the shallow areas were frozen. He found no difference in diving abilities in the sexes, with the longest observed dive (47 seconds) made by a female. Olney and Mills (1963) stated that most foraging occurs at depths less than 4 meters, and rarely do the birds exceed 9 meters in their dives. They also noted that goldeneyes often turn over stones with their bills while underwater, searching for aquatic insects or other organisms.

Sociality, densities, territoriality. Carter (1958) estimated that about 150 pairs (five-year average) of goldeneyes nested yearly on a New Brunswick study area containing 16,000 acres of hardwood swamp, representing about one pair per 100 acres of swamp. Usually the available nest sites in the form of natural cavities are well scattered and result in a rather sparse, randomly distributed breeding pattern for this species. Grenquist (1963) reported how the goldeneye population in Finland underwent a rapid increase in the 1950s following a program of adding artificial nesting boxes, but that as the population increased and the availability of nesting boxes became limited, there was increased conflict among females for the boxes, resulting in a high incidence of unhatched eggs as well as serious fighting among the females. If nest sites are available, females will sometimes nest very close to one another (Siren, 1957a).

Interspecific relationships. In North America, the wood duck is the only hole-nesting species that extensively overlaps in breeding range and nesting requirements with the common goldeneye. Prince (1968) has analyzed their differences in nest site characteristics and noted several criteria that might contribute to a reduction in competition for such sites. Hooded mergansers possibly compete locally with goldeneyes for available nests. Common mergansers sometimes also nest in cavities, and Grenquist (1963) reported that when females of both species used the same nesting box, it was deserted by the goldeneye. Red-breasted mergansers can be either surface- or cavity-nesters, but cavities they use are generally at ground level.

General activity patterns and movements. Goldeneyes are well known to be daytime feeders, often "rafting" in deeper waters at night. Breckenridge (1953) studied one such raft of wintering birds on the Mississippi

River near Minneapolis, Minnesota. He found that the evening flight to the rafting area began about an hour before sunset and lasted until about an hour after sunset, or when it was virtually dark. Birds were found to move into this area from as far as 27 miles upstream and 10 miles downstream. The maximum size of the raft was found to be somewhat over 600 birds. Such rafts no doubt help maintain ice-free areas of water.

Carter (1958) noted that spring migrant goldeneye flocks were usually found in fairly large flocks that remained rafted on deeper waters, while the resident birds occurred in pairs or small groups well apart from these flocks until later in the day, when they foraged on the outskirts of the transient flocks. He found little foraging to occur during the earliest hours of daylight, and after the midday foraging period, shortly before dusk, the paired birds once again separated from the others.

Social and Sexual Behavior

Flocking behavior. Breckenridge (1953) found that the average flock size of the wintering goldeneyes he studied on the Mississippi River dropped from 32 to 2.7 birds between December and the end of March, apparently reflecting the gradual pairing of birds. However, active courtship was seen throughout the entire period, without any definite peak. Probably some display continues among paired birds, since Carter (1958) mentions seeing "courtship" among both paired and unpaired groups.

Evidently there is a tendency for the older and the paired birds to migrate northward more rapidly than the younger ones or the unpaired adults. Carter reported this to be the case in New Brunswick, and observations in central Washington (Johnsgard and Buss, 1956) support this. A locational segregation of paired and unpaired birds was also supported by the latter study.

Pair-forming behavior. The social displays of male common goldeneyes are probably more diverse and complex than those of any other North American waterfowl and cannot be adequately summarized here. Several studies (e.g., Myres, 1959a; Dane and Van der Kloot, 1964) have dealt with these complexities in considerable detail. As mentioned earlier, social display lasts over a period of several months, from early winter through spring with a probable peak in March. Carter (1958) reported that by the time the goldeneyes arrive in New Brunswick in late March, nearly 80 percent of the adult birds are already paired. A later arrival of immature birds causes a reduction in this percentage.

Social display usually involves a small group of birds; Nilsson (1969) reported a typical group as consisting of 2 females and 3 to 5 males, and a maximum of 5 females and 20 males. Aggressive situations stimulate display, and probably most of the male display postures and calls are products of ritualized aggression. These include a simple head-throw, a fast and slow form of head-throw-kick, a "bowsprit" and "masthead" posture, and a number of other less complex movements. Most of these are performed in a stereotyped manner having remarkable time-constancy characteristics (Dane and Van der Kloot, 1964; Dane et al. 1959).

Primary female postures and calls involve a highly ritualized inciting movement (also called "head-forward" and "jiving") and a distinctive neck-dipping movement that seemingly is a strong sexual stimulus to males. A weak screeching note accompanies this movement, whereas inciting is performed silently. The

Male displays of the common goldeneye, including (1–2) two phases of fast head-throw-kick, (3–4) two phases of bowsprit, and (5–8) four phases of slow head-throw-kick.

Common goldeneye copulation sequence (1–8), including male (1) drinking, (2) wing-stretching, (3) jabbing, (4) preening-behind-the-wing, (5) steaming to female, (6) treading, and (7–8) postcopulatory rotations.

Common goldeneye, male head-throwing and female

female will often follow a favored male while performing inciting movements, to which the male usually responds with lateral head-turning, while swimming ahead of her (Johnsgard, 1965).

Copulatory behavior. Behavior patterns associated with copulation are nearly as complex as those associated with social courtship. The female typically assumes a prone position, often after mutual drinking movements by the pair, and remains in it for a prolonged period as the male performs his precopulatory displays. These include a large number of seemingly unritualized comfort movements. The most common of these is display-drinking (Nilsson, 1969; Lind, 1959), with a stretching of the wing and leg of secondary frequency, and a number of other movements such as bathing, dipping, shaking the bill in the water ("water-twitching"), and rolling the cheeks on the shoulders. Just prior to mounting, the male performs a vigorous series of water-twitching ("jabbing") movements, preens suddenly and momentarily, and immediately approaches the female in a "steaming" fashion. During treading the male normally shakes his wings one or more times, and before releasing the female he pulls her around in a partial or complete circle ("rotations"). He then swims directly away, with his head feathers fluffed, performing lateral head-turning movements while uttering low grunting sounds (Johnsgard, 1965).

Common goldeneye, male head-throwing and female head-pumping

Nesting and brooding behavior. In New Brunswick, Carter (1958) found that females begin looking for nesting sites about two or three weeks after their arrival, and the first eggs are laid about a week after that. If previously used nest sites are still available from past years, these are often used, sometimes up to five years in succession, but this is not invariable (Siren, 1957a). No down is deposited prior to the first egg, but by the time the clutch is complete the eggs are usually well surrounded by down. More down is added during the first week or two of incubation (Carter, 1958).

Normally on the morning following hatching, the female calls the young from the nest, and they typically jump out of the cavity in rapid succession. The complete evacuation of the nest by the brood might occur very rapidly; in five cases the range was 40 to 150 seconds. After all the young have left the nest, the family usually rests a few minutes and then begins to move toward water (Bauer and Glutz, 1969). At times, the newly hatched young must walk a mile or more before reaching water (Carter, 1958). Swimming females have been seen letting the young climb up on their backs to rest, but the carrying of young on the back while in flight is still considered highly questionable (Johnsgard and Kear, 1968).

Studies in Finland indicate that, after hatching, the adult female usually leads her young to small forested pools, often abandoning one or more of her ducklings in such pools and moving to another, so that by the time the young are fledged she might be caring for only a single offspring (Siren, 1957b). Carter (1958) reported seeing untended young quite often; such birds constituted about 14 percent of the total ducklings seen. He also noted that females seemed to abandon their entire brood at a much earlier age than do most species of ducks.

Postbreeding behavior. In New Brunswick, males remain in the general breeding area for a time after abandoning their mates and congregate in small groups. Most of them evidently do move to river mouths and coastal inlets before becoming flightless. Females are by then abandoning their broods and begin to molt about the time the young are nearly fledged. Females evidently remain in the general area of breeding to complete their molt. There is a gradual movement of juveniles toward the coast as they fledge, and there they gather with older birds in small groups on deep water, foraging in shallower waters during early morning and evening hours (Carter, 1958).

Hooded Merganser
Lophodytes cucullatus Linnaeus 1758

Other vernacular names. Fish duck, hairyhead, sawbill

Range. Breeds from southeastern Alaska and adjacent Canada eastward through the southern and middle wooded portions of the border provinces to New Brunswick and Nova Scotia; southward to Oregon and Idaho, in a southeasterly direction across the wooded parts of the northern Great Plains to the Mississippi Valley, and from there to the Atlantic coast and sporadically as far south as the Gulf Coast states. Winters along the Pacific coast from Mexico north to southern British Columbia, along the Gulf Coast, along the Atlantic coast north to the New England states, and to some extent in the interior, especially on the Great Lakes.

Subspecies. None recognized.

Measurements. *Folded wing:* Delacour (1959): Males 195–201 mm, females 184–198 mm. Palmer (1976): 12 males, ave. 198.5 mm; 12 females, ave. 185.2 mm.
   Culmen: Delacour (1959): Males 38–41 mm, females 35–39 mm. Palmer (1976): 12 males, ave. 39.5 mm; 12 females, ave. 38.3 mm.

Weights (mass). Nelson and Martin (1953): 24 males ave. 1.5 lb. (679 g), max. 2.0 lb. (907 g); 20 females ave. 1.2 lb. (543 g), max. 1.5 lb. (679 g). Kortright (1942): 19 adult males ave. 680 g; 12 adult females ave. 554 g.

Identification

In the hand. Apart from the extralimital Eurasian smew, this is the only sea duck with a merganser-like bill (narrow, serrated, with a large, curved nail), a culmen length less than 45 mm, and a folded wing measurement less than 205 mm. Additionally, the rounded crest, yellowish legs, and ornamental black or brownish tertials with narrow white or ashy stripes are all distinctive.

In the field. On the water, both sexes appear as small ducks with long, thin bills and fanlike crests that are usually only partially opened. Only the bufflehead has a comparable white crest, and that species lacks reddish brown flanks, has no black margin on the crest, and has a much shorter bill. Immature males or females appear as slim grayish brown birds with a brownish head and a cinnamon-tinted crest.
   In flight, hooded mergansers lower the crest and hold the head at the same level as the body, making a streamlined profile, and exhibit their distinctive black and white upper wing pattern and an underwing

coloration that is mostly silvery gray and whitish. Females are not notably vocal, but during courtship the male's distinctive low-pitched call, a rolling, froglike *crrrooooo*, can be heard over a considerable distance.

Age and Sex Criteria

Sex determination. The presence of mostly pale gray middle and lesser coverts indicates an adult male, but either sex can have brownish black or brownish gray coverts. Some immature males can be recognized by having one or more pale gray feathers among the surrounding dark coverts, but other wing criteria are apparently unsuited for sexing young birds (Carney, 1964). Until the appearance of white crest markings, which normally occurs before the male is a year old, young males cannot be readily distinguished from females.

Age determination. Males can be readily aged by the fact that first-year males lack pale gray middle and lesser upper wing-coverts, or at most might have only a few feathers of such coloration. Females in their first year can be distinguished from older birds by their duller, browner overall coloration and their undeveloped crests (Bent, 1925).

Distribution and Habitat

Breeding distribution and habitat. This strictly North American species has a breeding range somewhat similar to that of the wood duck, with eastern and western segments that are seemingly well isolated from one another. In western North America the hooded merganser breeds as far north as southeastern Alaska, with young having been seen on the Stikine, Chilkat, and Innoko rivers (Gabrielson and Lincoln, 1959). The range continues into British Columbia, including the Queen Charlotte Islands, and the mountains of western Alberta (Godfrey, 1986).

In Washington the hooded merganser breeds at least in the northern part of the state (Yocom, 1951), and its western breeding range apparently extends to Oregon, where it is a locally uncommon summer resident west of the Cascades, and at middle altitudes in the Cascades (Gilligan et al., 1994). It historically nested in California south to Yosemite National Park. Thanks to a recent nest-box program, it now breeds locally in northern California (Pandolfino, Kwolek, and Kreitinger, 2006) and possibly also breeds locally in Nevada (Stallcup, 2002). Along the Rocky Mountains the hooded merganser extends southward locally through eastern Idaho and western Montana (Marks, Hendricks, and Casey, 2016). There are three breeding records, the most recent in 2006, for Wyoming (Faulkner, 2010).

It is rather uncertain whether any breeding occurs in Saskatchewan, but breeding birds have been found in the southern and middle wooded portions of Manitoba as well as eastward through the comparable portions of Ontario, Quebec, and New Brunswick. Rare breeding in Nova Scotia evidently marks the eastern limits of the Canadian range (Godfrey, 1986).

South of Canada, local or sporadic breeding has been reported from Texas, Oklahoma, Missouri, Kansas, Nebraska, South Dakota, and North Dakota, and more widespread breeding occurs in Missouri, Iowa,

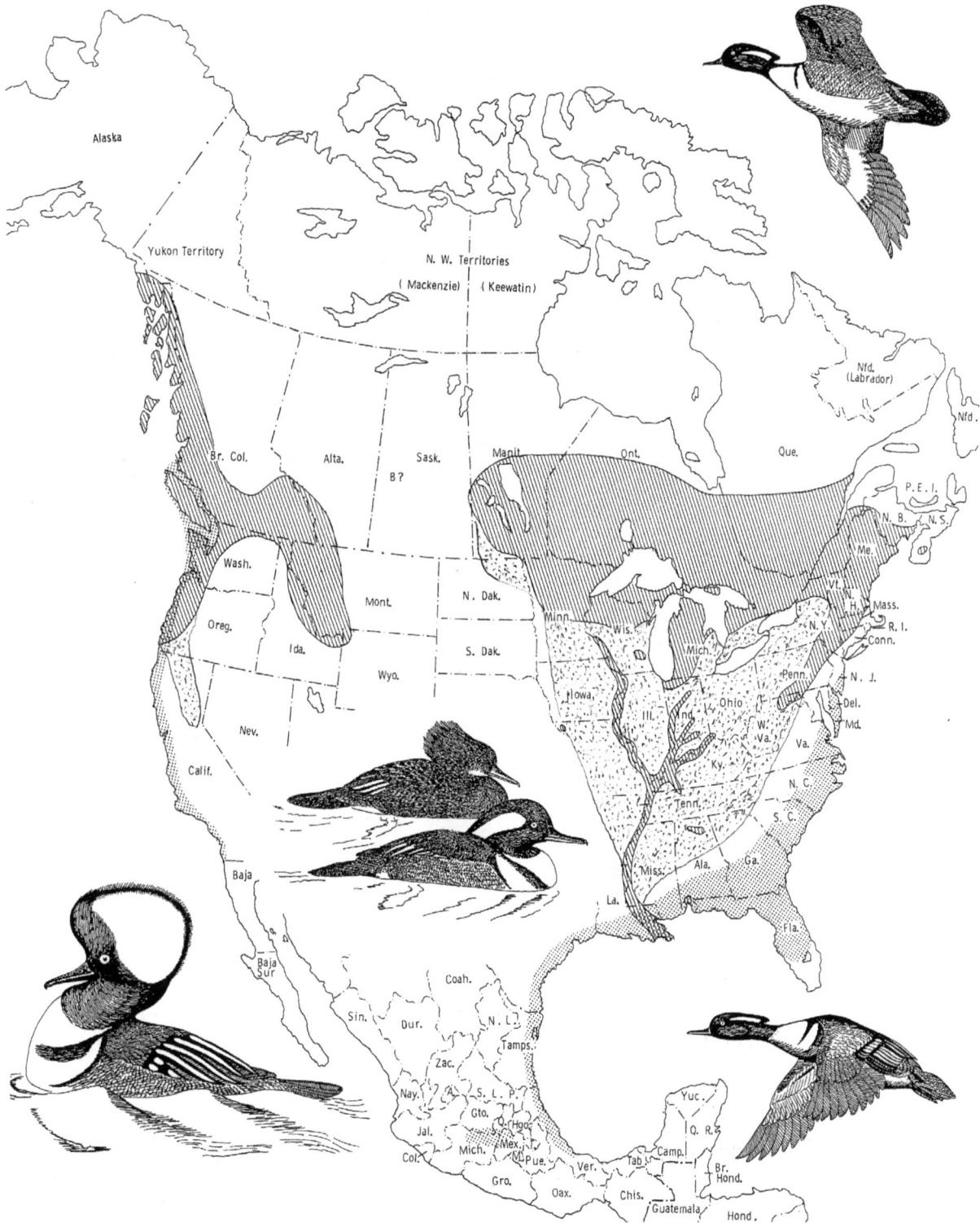

The North American breeding (hatched) and wintering (shaded) ranges of the hooded merganser. Regions of newly acquired or previously unrecognized breeding are shown by stippling.

Hooded merganser, pair in breeding plumage

Minnesota, Wisconsin, Michigan, Indiana, New York, Vermont, New Hampshire, and Maine. Breeding also occurs along the length of the Mississippi and Ohio rivers and their wooded tributaries. Local or sporadic breeding also extends south along the Atlantic coastal states from Massachusetts to Florida.

The preferred breeding habitat of the hooded merganser consists of wooded, clear-water streams and, to a lesser degree, the wooded shorelines of lakes. The combination of food in the form of small fish and invertebrates in water sufficiently clear for foraging and suitable nest sites in the form of tree cavities is probably a major factor influencing its breeding distribution. Like the wood duck, it seems rather sensitive to cold, and its breeding range is considerably more southerly than are those of the two other mergansers.

Wintering distribution and habitat. The western segment of the hooded merganser population primarily winters along the Pacific coast, from as far north as southern British Columbia (Godfrey, 1986) to southern

Hooded merganser, male and female

California and occasionally reaching Baja California (Leopold, 1959). Small numbers are sometimes found on the southern Great Lakes during winter, but most of the eastern wintering population can be found from Massachusetts southward along the Atlantic coast to northern Florida, and along the Gulf Coast, with occasional birds reaching Tamaulipas and Veracruz (Leopold, 1959). In the Chesapeake Bay region only small numbers regularly winter, and these are mostly in the fresh and brackish estuarine bay marshes, with some usage of river bottomlands and fresh estuarine bays or interior impoundments. Salt estuaries and open ocean are evidently avoided (Stewart, 1962).

General Biology

Age at maturity. Females are evidently sexually mature and breed when two years old, in the third spring of life (Morse et al., 1969).

Pair-bond pattern. Pair-bonds are renewed yearly, with an associated period of social display (Johnsgard, 1961a). Pair-bonds are broken when the female begins incubation (Morse et al., 1969).

Nest location. McGilvrey (1966) reported that six of eight nesting boxes used by hooded mergansers were in open impoundments rather than in impoundments with dead timber present. Morse et al. (1969) found that boxes closely adjacent to water were much more heavily used than those some distance from it. Minimum and optimum size criteria for natural cavities have not yet been reported but probably would tend to be smaller than those used by wood ducks. If the same habitat characteristics apply to nesting as to brooding, then the findings of Kitchen and Hunt (1969) might be important. They found greatest brood usage on rivers with high food resources, rivers with relatively fast currents, wide rather than narrow rivers, and those with moderately deep channels. Of 65 brood observations cited in the literature, 46 (71 percent) were on rivers or river-related habitats, 12 were on beaver ponds, and 7 were on other standing-water habitats.

There is a high incidence of nest-site fidelity, with 33 to 89 percent of females returning to a group of nest boxes in subsequent years over a five-year period, although different boxes were sometimes used (Hansen, 1981). Dugger, Dugger, and Fredrickson (1999) established that some females (3.8–10.8 percent at two different sites) returned to their natal nest boxes for their breeding.

Clutch size. Morse et al. (1969) reported an average clutch size of 10.2 eggs for 55 nests, with a range of 6 to 15. Clutch sizes tended to decrease with the season and were smaller (average 9.4 eggs) for eight initial breeders than for ten older breeders (average 10.8 eggs). The egg-laying rate averaged one per 48 hours. The eggs of hooded mergansers are nearly round (width:length ratio 1:1.24), a trait typical of many cavity-nesting birds that allows more eggs to fit into a confined space. The eggs are also relatively thick-shelled, making them more crack resistant and less likely to break in large clutches. This trait is typical of many avian brood parasites (Johnsgard, 1997).

Brood parasitism (or dump-nesting), the deposition of eggs into the nests of other birds, is common in hooded mergansers. Zikus (1990) judged that clutches of more than 13 hooded merganser eggs in a nest were the result of eggs from other females, a condition he found in 45 percent of observed clutches. He thus believed that up to 75 percent of the nests he studied in Minnesota were the result of multiple female input, and that such conspecific brood parasitism is adaptive. However, nest success is reduced in very large clutches. Dugger, Dugger, and Fredrickson (2009) reported that unparasitized hooded merganser nests had hatching success rates of 70.6 percent, whereas those that had been conspecifically parasitized had hatching rates of 59 percent. Nest parasitism by wood ducks reduced the rate of nest success even more than did conspecific parasitism.

Incubation period. The mean incubation period was 32.6 days for naturally incubated nests, with an observed range of 29 to 37 days (Morse et al., 1969). Other studies have found averages ranging from 29 to 33 days (Dugger, Dugger, and Fredrickson, 1996).

Fledging period. Estimated by McGilvrey (1966) to be about 70 days, a surprisingly long fledging period for such a small duck, and about 10 days longer than that of the red-breasted merganser, but about the same as that of the similar-sized and cavity-nesting Eurasian smew.

Nest and egg losses. Morse et al. (1969) found that eggs laid in 44 of 55 nests were successfully hatched, a nesting success of 80 percent. In these successful nests, 92.2 percent of the eggs hatched, and the average brood size at the time of hatching was 9.6 ducklings. Dump-nesting (or brood parasitism) caused some losses in Morse's study; in 1968 there were 6 such clutches among a total of 24, 3 of which were unsuccessful. However, Bouvier (1974) reported that four of five hooded merganser eggs hatched when present in common goldeneye nests, and 14 of 17 common goldeneye eggs hatched when in nests of hooded mergansers.

Dugger, Dugger, and Fredrickson (1994) reported that average nest success in nonparasitized and conspecifically parasitized hooded merganser nests was similar (70.6 versus 70.0 percent), but both were higher than in nests parasitized by wood ducks (61.4 percent), Additionally, hatching success for nonparasitized nests was 75 percent, versus 59 percent for conspecifically parasitized nests and 50 percent for nests parasitized by wood ducks.

In addition to competition from other hole-nesting ducks, there are many egg predators of hooded merganser nests, including mink, raccoons, black bears, and snakes, especially rat snakes (*Elaphe obsoleta*). Old pileated woodpecker (*Dryocopus pileatus*) nests are a favorite type of nesting cavity wherever they occur.

Juvenile mortality. Duckling losses sometimes appear to be quite high in this species (McGilvrey, 1966), although the sources of such mortality can only be guessed at because of nest parasitism effects on brood sizes. Postfledging mortality rates of juveniles are still unreported.

Adult mortality. Dugger, Dugger, and Fredrickson (1999) estimated an annual survival rate of nesting females in Missouri as 66 percent, but there were marked annual variations from 42 to 100 percent over 11 years. Morse et al. (1969) noted that 11 of 18 adult females banded in 1966 and 1967 returned to nest the following year, representing a minimum annual survival rate of about 60 percent. Of the 11 returning females, 7 used the same box the following year, and all nested within three miles of their previous nest site. Thus, like other hole-nesting species that have been studied, there is a high level of nest-site fidelity, and the numbers of returning females might provide a reasonable basis for estimating minimum adult annual survival rates. The greatest known longevity was a female both banded in Nova Scotia and recovered there 11 years and 3 months later (Baldassarre, 2014).

General Ecology

Food and foraging. Surprisingly little is known of the foods of this species. Bent (1923) indicated that insects make up a large part of the food, together with small fish, frogs, tadpoles, snails, other mollusks, crayfish, and other small crustaceans. It forages on both muddy and stony bottoms and consumes a rather small

Hooded merganser, male calling and crest-raising

amount of plant materials. Stewart (1962) reported that all of the ten hooded mergansers he obtained in the Chesapeake Bay area had been feeding on various fish. Crustaceans (mud crabs and crayfish) and insects (caddisfly larvae and dragonfly naiads) were also found in some of these birds.

In a summary of ten food consumption studies, Dugger, Dugger, and Fredrickson (1994) reported that fish remains were present in nine, insects in five, crabs and/or crayfish in four, and frogs in two. Centrarchid fish (bass and sunfish) were present in two of the studies, as were dragonflies and caddisflies. Plant materials were also present in seven, including acorns in one.

Relatively clear waters with sandy or cobble bottoms are preferred over mud-bottom habitats for foraging. Kitchen and Hunt (1969) found a preference among females and broods for foraging in fairly fast-moving waters, in waters having an average depth of only 20 inches, and cobble-bottom stream areas rather than mucky areas. Such areas tend to be rich in fish, crustaceans, and aquatic insects, although the specific foods taken were not determined.

Hooded merganser, male crest-raising

Sociality, densities, territoriality. Except where nesting boxes are present, the relative rarity of suitable nesting cavities possibly places a limit on maximum breeding densities in an area. However, Kitchen and Hunt found no lack of suitable cavities in their study. Instead, they correlated breeding density (broods observed) with river characteristics related to food availability. The highest brood use figure they obtained was 2.14 broods per mile of river. Additionally, heavily wooded rivers were favored over brush-lined rivers, and marshy rivers had the lowest brood densities.

Interspecific relationships. The relationships of nesting hooded mergansers and wood ducks in their overlapping areas of breeding have not yet been analyzed but would be of considerable interest. The fact that hooded mergansers prefer habitats with rapidly flowing water over still waters, whereas wood ducks prefer slow-moving rivers and ponds, would tend to reduce competition for nest sites, although mixed clutches in areas of overlap are not uncommon. Additionally, different food sources would certainly influence the local distribution of pairs and families.

Nest parasitism by wood ducks is probably an important interspecific interaction in the southern states, with a reported 42 percent of hooded merganser nests in southwestern Missouri being affected (Dugger, Dugger, and Fredrickson, 1994). Farther north, hooded mergansers are likely to lay their eggs in the nests of common goldeneyes, and vice versa. In Minnesota, Bouvier (1974) reported that 4 out of 5 hooded merganser eggs that had been deposited in common goldeneye nests hatched, as did 14 of 17 eggs of common goldeneyes that were laid in hooded merganser nests. Common merganser eggs have also rarely been found in hooded merganser clutches, and Phillips (1926) mentioned that in Maine barred owl eggs have been found among those of hooded mergansers.

Phillips also indicated that the hooded merganser rarely mixes with the larger species of merganser and forages in different habitats from them. It is also less dependent than the other two mergansers on fish, relying to a greater extent on insects and crustaceans. It sometimes forages in the company of buffleheads, and no doubt the two species feed to some extent on the same kinds of invertebrates.

General activity patterns and movements. Hooded mergansers are daylight foragers; indeed they probably require both good light and clear water to catch such active prey as fish. How much time per day is spent in foraging has not yet been reported, but incubating females normally leave their nests three times a day for this purpose (Morse et al., 1969), in early morning, midday, and late afternoon.

Social and Sexual Behavior

Flocking behavior. Large flocks are not typical of this species; most writers (such as those cited by Bent, 1923) report that 5 to 15 birds typically constitute a flock. Flocks as large as 100 birds have been seen (Harper, in Phillips, 1926) but are very unusual.

Pair-forming behavior. Pair-forming displays by wild birds have been reported by so few observers that it is difficult to judge when most pair formation does occur. Harper (in Phillips, 1926) observed active display in early February among wild birds, and I have seen it among captive individuals (Johnsgard, 1961a) throughout the winter and spring months. Harper noted that courting flocks contained 3 to 10 birds including 1 to 3 females. Male displays of the hooded merganser are in large measure related to the ornamental crest. Crest-raising, either independently or in conjunction with other displays, is very frequent. The head and erect crest are often shaken laterally, as the bird rises slightly in the water; such shaking often precedes a head-throwing movement that includes a rolling froglike note. A silent, elliptical neck-stretching, or "pumping," movement is also frequent and is seemingly hostile in function. A turning of the depressed crest toward a female, diagonal tail-cocking, body-shaking ("upward-stretch"), and wing-flapping are all relatively frequent during pair-formation activities and all appear to represent displays (Johnsgard, 1961a).

The female's movements include a pumping movement, similar to the male's and often performed simultaneously with the male's display, and a variation of inciting ("bobbing") that is apparently rather rare

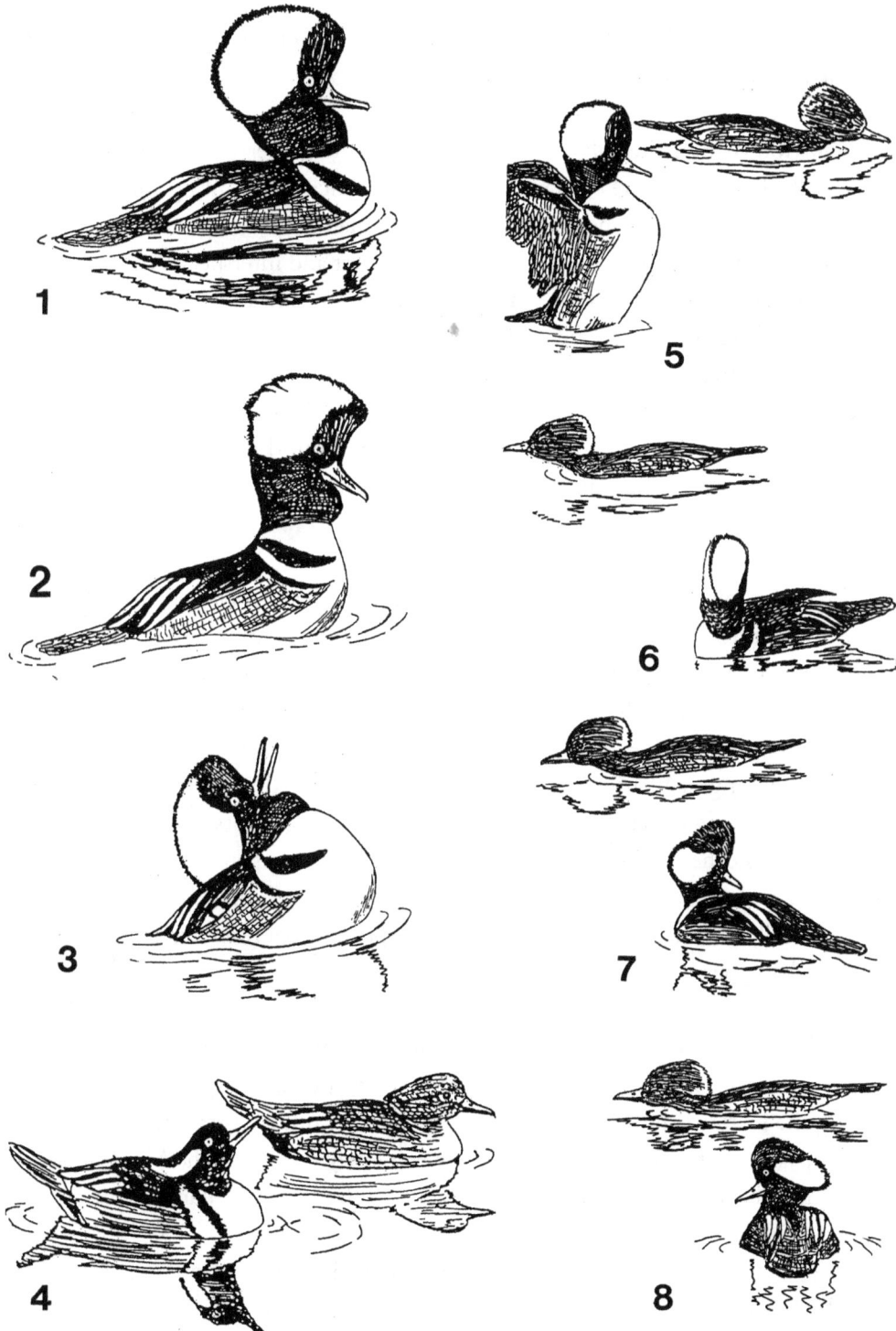

Male displays of hooded merganser, including (1–3) head-throw sequence, and (4–8) precopulatory sequence, including (4) drinking (5) wing-flapping, (6) preening-behind-the-wing, and (7–8) tacking approach to female.

in this species. As in other ducks, the usual response of the male to inciting is to swim ahead of the female and turn the back of his head toward her.

Copulatory behavior. As with the goldeneyes and the other mergansers, the female hooded merganser assumes an outstretched prone posture on the water well in advance of mounting and often after the pair has performed ritualized drinking movements. At this time the male begins to perform almost continuous and rather jerky back-and-forth head movements with lowered crest and intersperses these with drinking movements and body shakes. Suddenly he begins a number of vigorous head-shakes with his bill in the water, stops, and performs a body-shake or a few wing-flaps, preens once in the region of his back, and starts toward the female. He approaches her in a somewhat zigzag fashion, presenting first one side, then the other, of his raised or nearly raised crest toward her. He then mounts the female, and during treading flicks his folded wings. Before releasing the female's nape he pulls her around in a partial rotary movement. After treading, he swims rapidly away with an erect crest, terminating this swim either with a quick dive or with bathing (Johnsgard, 1961a).

Nesting and brooding behavior. After locating suitable nest sites, females begin to deposit their eggs at the rate of one every other day. Unlike many ducks, this species evidently does not normally deposit down in the nest until the initiation of incubation. Males desert their mates at about this time, and after incubation is under way the female usually leaves her nest only three times a day, in early morning, midday, and late afternoon. Several writers have commented on the secretive manner of the hen when returning to her nest. After a relatively long incubation period of 32 to 33 days, the ducklings hatch. They usually remain in the nest a full day, leaving the cavity early the following morning (Morse et al., 1969).

The female might tend her brood near the hatching area or move them into other water areas, but evidently seeks out waters less than 20 inches deep that are quite close to timber (McGilvrey, 1966). At what stage the female usually abandons her brood to begin her postnuptial molt apparently has not been determined, but the fledging period in this species is unusually long.

Postbreeding behavior. The males probably begin to molt fairly soon after deserting their mates, but few observations on the behavior and movements of males in the postnesting season are available. They evidently become quite secretive and probably move into heavily timbered streams to complete their flightless period. Phillips (1926) noted that adult males are extremely rare along the Massachusetts coast during fall, leading him to believe that they perhaps migrate by a different route than do females and immatures, although there is no current evidence for that.

Estimates of total fall North American populations have varied widely, from as few as 76,000 (Bellrose, 1976) to as high as 275,000 to 350,000 (Dugger, Dugger, and Fredrickson, 1994).

Red-breasted Merganser
Mergus serrator Linnaeus 1758

Other vernacular names. Fish duck, saw-bill

Range. Breeds in Greenland, Iceland, the British Isles, northern Europe and Asia from Scandinavia to Kamchatka, the Aleutian Islands, and from Alaska eastward across nearly all of Arctic Canada except the northern part of Nunavut and the Arctic islands, south to northern British Columbia and Alberta, central Saskatchewan and Manitoba, southern Ontario, the Great Lakes states, northern New York, New England, and the Maritime Provinces east to Newfoundland. Winters mostly on salt water, in North America from southeastern Alaska south to Baja California, the Gulf Coast, the Atlantic coast from Florida to the Gulf of St. Lawrence, and inland in smaller numbers as far north as the Great Lakes.

North American subspecies. *M. s. serrator* L. Common Red-breasted Merganser. Breeds as indicated above, except in Greenland.

M. s. schiøleri Salomonsen. Greenland Red-breasted Merganser. Resident in Greenland. Not reported from mainland North America.

Measurements. *Folded wing:* Delacour (1959): Males 224–260 mm, females 217–230 mm. Palmer (1976): 12 ad. males 238–257 mm, ave. 248.5 mm; 12 ad. females 213–239 mm, ave. 224 mm.

Culmen: Delacour (1959): Males 53–62 mm, females 48–55 mm. Palmer (1976): 12 ad. males 57–60 mm, ave. 59 mm; 12 ad. females 50–58 mm, ave. 55 mm.

Weights (mass). Nelson and Martin (1953): 18 males ave. 2.5 lb. (1,133 g), max. 2.9 lb. (1,314 g); 17 females ave. 2.0 lb. (907 g), max. 2.8 lb. (1,268 g). Schiøler (1926): 10 winter adult *M. s. schiøleri* males ave. 1,209 g (2.67 lb.); 10 adult females ave. 959 g (2.12 lb.).

Identification

In the hand. The long, narrow, serrated bill with a hooked tip will distinguish this species from all other mergansers except the common merganser. In the red-breasted merganser the bill is distinctive in that (1) the nostrils are located in the basal third of the bill, (2) the feathering on the side of the upper mandible reaches considerably farther forward than that on the lower mandible, (3) the upper mandible is relatively longer and lower at the base than in other mergansers, at least six times as long as it is high at the base when measured from the cutting edge to the highest unfeathered point, and (4) the bill has a smaller, narrower nail at the tip. Both sexes are smaller than the common merganser: adult males and females have maximum folded wing lengths of 260 and 230 mm, respectively, versus minimums of 269 and 246 mm in common mergansers.

In the field. When in nuptial plumage, the male red-breasted merganser can be recognized by its green head, which extends backward into a shaggy double crest and is separated in front from a brownish breast by a white foreneck. The sides and flanks appear to be a light gray, bordered anteriorly with a black patch having regular white spots. The female is not nearly so "two-toned" as the female common merganser; her grayish body merges gradually with the brownish head, and neither its paler throat nor its lores are in strong contrast to the rest of the head. The female calls of the two species are very similar, but the courtship notes of the male red-breasted merganser are a somewhat catlike *yeow-yeow*, uttered during its strange posturing. In flight, both sexes resemble the common merganser, but males exhibit a brownish breast band, while females appear to have a darker brown, less reddish head and neck color, which gradually merges with the grayish breast and pale underparts.

Age and Sex Criteria

Sex determination. In adults, white middle and lesser coverts, and tertials that are either black or white margined with black indicate a male. First-year males begin to acquire male-like features about December, when black feathers appear on the head, mantle, and scapulars, whereas the white scapular feathers do not appear until the end of March.

Age determination. First-year males can be readily aged by their mostly grayish black tertials, which are narrow and have wispy tips, and the absence of immaculate white on the middle and lesser coverts. Adult females can be distinguished from first-year birds by having tertials and greater tertial coverts that are smoothly rounded, rather than narrow with wispy tips (Carney, 1964).

Distribution and Habitat

Breeding distribution and habitat. The North American breeding distribution of this species is the most northerly and most extensive of any of the merganser species. It breeds in the Aleutian Islands from Attu to the Alaska Peninsula as well as on Kodiak Island (Murie, 1959) and probably also on St. Lawrence Island (Fay, 1961). On the mainland of Alaska it has a wide occurrence throughout most of the state, although it is less frequent and perhaps is only an occasional breeder from Kotzebue Sound north and east along the Arctic coast (Gabrielson and Lincoln, 1959).

In Canada it breeds from the Arctic coast of Yukon Territories and through the Mackenzie District of Northwest Territories eastward across most or all of mainland Nunavut, and southward to northern British Columbia, northern Alberta, central Saskatchewan, and to probably all of the Maritime Provinces. It also breeds on southeastern Victoria Island, and southern, and locally northern, Baffin Island, as well as along the southeastern and southwestern coasts of Greenland (Titman, 1999).

South of Canada, it breeds locally in northern Minnesota (Lee et al., 1964a), uncommonly in northern Wisconsin (Jahn and Hunt, 1964), in northern Michigan (Brewer, McPeek, and Adams, 1991), and locally

The North American breeding (hatched) and wintering (shaded) ranges of the red-breasted merganser. Regions of newly acquired or previously unrecognized breeding are shown by stippling.

in the northeastern states to Vermont (rarely) and Maine (Titman, 1999). There have been a few isolated records of breeding farther south, such as in Pennsylvania, North Carolina, and South Carolina. The favored breeding habitat would seem to be inland freshwater lakes and streams that are not far from the coast. Deep, rock-lined lakes are seemingly favored over tundra ponds (Snyder, 1957), but the ground-nesting adaptations of this species allow it to nest in nonforested situations well away from a source of hollow trees.

In northern Europe this species nests primarily in lakes and rivers having barren shores and clear waters, either among forests or in tundra zones. Locations having many available cavities, such as boulder-strewn areas, are rich in potential nest sites and thus tend to be favored (Hildén, 1964).

Wintering distribution and habitat. In Alaska, red-breasted mergansers occur widely in the Aleutian Islands (Murie, 1959), and along the southern and southeastern coasts of Alaska they are also fairly common (Gabrielson and Lincoln, 1959). In Canada they winter along the entire coast of British Columbia, occur in small numbers on the Great Lakes, and extend from the St. Lawrence Valley to Newfoundland and the other Maritime Provinces (Godfrey, 1986).

South of Canada, the red-breasted merganser winters along the Pacific coast from Puget Sound southward through Oregon and California to Mexico, where it is the commonest of the wintering mergansers (Leopold, 1959). Leopold reported seeing it along the Pacific coast and in the central highlands of northern Mexico but not along the Caribbean coastline.

Along the Atlantic coastline, red-breasted mergansers are prevalent during winter from Maine southward at least as far as Georgia (Burleigh, 1958). They occur uncommonly along the Gulf coast of Florida (Chamberlain, 1960), extending west through Louisiana and coastal Texas. In the Chesapeake Bay region they are frequent during the winter months, although larger numbers pass through on migration (Stewart, 1962).

Stewart reported that, in contrast to the freshwater tendencies of the common merganser, this species is characteristic of saline tidewaters, occurring all the way from open ocean through salt and brackish estuarine bays to fresh or slightly brackish waters, of which it makes only small usage. In marine habitats it tends to avoid deep or rough waters, preferring sheltered and relatively quiet areas where small fish are abundant and can be caught in shallow waters (Hildén, 1964).

General Biology

Age at maturity. Sexual maturity probably occurs in two years, judging from plumage sequences (Dementiev and Gladkov, 1967). Some sexual display has been observed in first-year males. However, only a few two-year-olds breed, and most females probably don't nest until their third year (Titman, 1999). In Iceland, Bengtson judged that 18 percent of adult females were nonbreeders.

Pair-bond pattern. Pair-bonds are renewed yearly, following a prolonged period of social display that begins on wintering areas. Pair-bonds are relatively loose, and many instances of apparent polyandry and polygyny have been noted (Bauer and Glutz, 1969).

Red-breasted merganser, male in breeding plumage

Nest location. Unlike the two other North American mergansers, this species regularly nests away from trees. In Iceland, where trees of substantial size are lacking, red-breasted mergansers nest commonly. Bengtson (1970) noted that 63 percent of the 332 nests he found there were in holes or cavities, while 60 nests were under high shrubs, 49 under low shrubs, and 14 under sedge or tall forb cover. Bengtson reported a distinct island-nesting tendency for this species, and a placement of nests close to water (modal distance category 10–30 meters). Hildén's (1964) study in the Valassaaret Islands of Finland included 238 nests. He found the nests were usually under boulders (39 percent), under dense juniper bushes (26 percent), or under *Hippophae* thickets (15 percent).

Unlike the common merganser, nesting occurs on both small islets and larger central islands, but the favored nesting substrate is associated with boulders. Islets with herbaceous vegetation are favored over those with grassy or wooded vegetation, apparently because of the presence of nesting lesser black-backed gulls (*Larus fuscus*). Curth (1954) reported that natural nesting cavities of this merganser averaged 27 cm wide and 7.8 cm deep. On the basis of studies of the use of nesting boxes, the preferred box dimensions were 17 to 23 cm high, 28 to 43 cm wide, and 42 to 50 cm long. The preferred entrance hole dimensions were 9–12.5 × 11.5–12.5 cm (Grenquist, 1958).

Clutch size. Hildén (1964) reported that 144 clutches averaged 9.23 eggs, with a range of 6–17 and a mode of 9. Bengtson (1971) found an average clutch of 9.5 eggs for 158 first clutches in Iceland. Curth (1954) reported average clutch sizes for of 9.8 and 9.9 eggs over two different years, excluding some abnormally large clutches that evidently resulted from dump-nesting. A sample of 27 second clutches in Iceland averaged 6.2 eggs (Bengtson, 1972). The egg-laying interval is 1.5 days (Curth, 1954).

Incubation period. Curth (1954) reported a mean incubation period of 31.8 days and an observed range of 29 to 35 days for wild birds. Slightly shorter periods have been estimated for artificially incubated eggs.

Fledging period. Apparently quite variable, one fledging period estimate was of 59 days (Bauer and Glutz, 1969) and others of 60 to 65 days (Cramp and Simmons, 1977) and 59 to 69 days (Kear, 2005).

Nest and egg losses. Hildén (1964) reported that among 67 clutches studied in a two-year period, 88 percent hatched, with six nests being taken by predators, one deserted, and one joint clutch with white-winged scoter eggs was unhatched. Although predators take many eggs during the laying period, predation losses are evidently low once incubation begins.

Curth (1954) reported high losses to predators, with as many as 30 percent of the nests being lost to common gulls (*Larus canus*). Ardamazkaja (cited in Bauer and Glutz, 1969) found that 91 percent of 790 nests near the Black Sea hatched during the years 1956–61, with yearly variations of 57.2 to 95.8 percent. Hildén (1964) rather surprisingly found that in spite of their cavity nesting, or otherwise hiding their nests well, both red-breasted and common mergansers suffered as high or higher losses of eggs to predators (mostly crows) than did surface-nesting ducks.

Juvenile mortality. Because of the prevalence of brood mergers in this species, average brood size counts are of little or no value in estimating prefledging losses. Hildén (1964) summarized past records of combined broods, which sometimes number from 30 to more than 60 ducklings in exceptional cases. Hildén's counts of total numbers of young in his study area over a three-year period indicated annual prefledging losses of 78, 84, and 92 percent. This seemingly high rate of juvenile mortality agreed well with an estimate of 86 percent for an earlier study.

Hildén correlated yearly brood success with weather and judged that bad weather was an important factor in brood survival. Further, broods in sheltered bays survived severe weather better than those on the outer archipelago. The relatively great agility of the young seemingly reduced losses to predatory gulls, although the great black-backed gull (*Larus marinus*) was nevertheless a dangerous enemy.

Red-breasted merganser, male in flight

Adult mortality. Estimates of annual adult mortality rates are apparently unavailable for this species. One New Brunswick female was still nesting at the age of at least 8 years (Titman, 1999), and individuals known to have survived at least 12 years have been reported in the United Kingdom (Kear, 2005).

Food and foraging. Foods of the red-breasted merganser in North America have not yet received as much attention as might be desired. Cottam and Uhler (1936) examined 130 stomachs from a variety of locations. They reported the following relative abundance of foods: minnows, killifish, and sticklebacks, 34 percent; commercial and game fish, 14 percent; carp and suckers, 3 percent; unidentified fish, 25 percent; and crustaceans or miscellaneous, 23 percent.

Munro and Clemens (1939) analyzed the foods of this species in British Columbia, where it is often considered a threat to salmon. On the basis of 77 specimens taken from November to January, it was found that the birds consumed primarily opaque salmon eggs, which are considered largely a waste product. Of 19 specimens taken from freshwater lakes and streams, sculpins (*Cottus*) were found in the largest number of stomachs (15), while salmonoid fry or fingerlings were present in only 3 specimens. Some sculpin eggs, insects, and annelid remains were also present. Among 15 specimens from salt water or estuaries, herring occurred with the greatest frequency (7 stomachs), sticklebacks were found in 3, sculpins in 2, and blennies and rock fish were present in one each. Crustacean remains were present in 5 specimens. Munro and Clemens believed that herring constitute the primary prey of red-breasted mergansers while in salt water.

Munro and Clemens noted that while feeding on the coast, red-breasted mergansers often swim close to shore in single file with their heads partly submerged. They are sometimes robbed by gulls, which often try to steal fish that the mergansers bring to the surface. An example of cooperative foraging described by Des Lauriers and Brattstrom (1965) involved seven birds swimming with their heads partly submerged in water less than 24 inches deep. They moved in a loose line, and when one began to chase a fish the others joined in to form a semicircle around the fish until it was finally caught.

Similar behavior has been noted in England (Hending et al., 1963) among flocks feeding in turbid waters. There, groups of 7 to 24 birds dove in near synchrony, with most of the birds submerging within 2 to 3 seconds. The average diving duration was 17.4 seconds, the average diving pause was 7.3 seconds, and the maximum observed diving duration was 29 seconds.

Sociality, densities, territoriality. Availability of nesting cavities or suitably dense shrub cover probably determines the nesting distribution of this species along with such factors as nesting gull colonies and island sites. Bengtson (1970) reported a higher nesting density for this species on islands than on the mainland of Iceland. Excluding cavity nests, he noted a density of 23 per square kilometer on islands, as compared with 10 per square kilometer on the mainland. Since nearly 70 percent of the nests he found were in holes and cavities, a considerably higher overall nesting density was evidently present.

Interspecific relationships. During winter there is probably little if any food competition between this species and the common merganser, since they tend to occupy saline and fresh waters, respectively. However, when both species are on fresh water, they seem to consume identical foods (Munro and Clemens, 1939).

A review by Mills (1962) suggests that the primary natural enemies of the red-breasted merganser are the great and lesser black-backed gulls, which consume both its eggs and ducklings. No doubt considerable destruction is caused by humans, either through the misguided control efforts by fishermen, or through the more pervasive and dangerous effects of pesticide and other chemical contaminant accumulations on the bird's reproductive efficiencies.

General activity patterns and movements. Red-breasted mergansers are necessarily daytime foragers, depending on their eyesight and underwater mobility for capturing prey. Nilsson (1965) reported that the

Red-breasted merganser, male

population he studied gathered each evening at a communal roosting place on a small islet. Courtship activity was common there during evening and early morning hours. By an hour after sunrise, they had spread out over the entire area. While foraging they were frequently harassed by scavenging gulls but were apparently always successful in evading them.

Social and Sexual Behavior

Flocking behavior. Red-breasted mergansers are relatively gregarious, and during migration as well as on wintering grounds they often occur in fairly large flocks. This no doubt is related in large part to the local concentrations of fish in suitable foraging areas, and perhaps also to the apparently greater efficiency of foraging in groups instead of individually. Munro and Clemens (1939) reported seeing flocks of 100 or more birds coming in to coastal British Columbia to forage on herring. Mills (1962) described winter flocks in Scotland of 30 to 400 birds, the latter groups apparently attracted to herring. With spring, these large flocks disperse toward their breeding grounds and no doubt gradually break up into pairs. Some summering flocks also occur locally in Scotland, which apparently represent molting accumulations.

Pair-forming behavior. Social display related to pair-formation begins on the wintering areas. Display takes on a highly distinctive form in this species, with the males often circling around the females and periodically performing a complex and rather bizarre series of movements called the "knicks" (German, meaning "bending"). Often, two males perform the display in synchrony or near synchrony, further adding to its rather bizarre appearance. From a resting posture, the male suddenly extends his neck and head diagonally upward, forming nearly a straight line ("salute"). After a momentary pause, he pulls the head downward toward the water, simultaneously gaping, uttering a faint catlike call, and raising the folded wings while tilting the tail downward ("curtsy"). The head is then retracted toward the shoulders, and the tail is more strongly down-tilted.

Occasionally a male will also suddenly dash over the water in a hunched "sprint" posture, throwing up a spray of water to both sides. Apart from the weak call associated with the knicks display, the birds are otherwise nearly silent, adding further strangeness to the almost unworldly activities. Females perform an infrequent but vigorous inciting movement, simultaneously uttering a harsh double note, but this display does not appear to be the primary stimulus for male display. Rather, it seems to prevent males from approaching the female too closely. A turning-of-the-back-of-the-head by males toward females has been reported by one observer (Curth, 1954), but in my experience this does not appear to play a significant role in social display (Johnsgard, 1965).

Copulatory behavior. Typically, copulation sequences are preceded by mutual drinking displays, followed by the assumption of a prone posture by the female. The male then performs a rather unpredictable series of drinking, preening, wing-flapping, and shaking movements. In my observations, the male always attempted to mount the female immediately after performing a rather rudimentary version of the knicks display (Johnsgard, 1965). After copulation, which may last up to as long as 13 seconds, the male rotates while still grasping the female, performs a knicks display, and both birds then begin bathing (Nilsson, 1965).

Nesting and brooding behavior. Females begin to look for nest sites as long as two or three weeks before egg-laying begins and are especially active during early morning hours. Brooding females usually leave their nests only for short periods of 62 to 125 minutes, and even shorter periods are typical during rainy weather. The times of such departures from the nest vary greatly but most often occur during the early morning hours. Apparently the females' food requirements are strongly reduced during incubation, and they will also drink salt water at such times. The young leave their nest site 12 to 24 hours after the hatching of the last duckling. The family might move 2 to 5 kilometers during their first few days, with the young sometimes riding on the back of the mother (Bauer and Glutz, 1969). Their cold hardiness is greater than that of dabbling ducks, pochards, or scoters, and similar to that of eiders.

Koskimies and Lahti (1964) found that three newly hatched red-breasted merganser ducklings retained their thermoregulation for at least three hours at a temperature of 0–2°C. However, prolonged periods of bad weather might greatly affect brood survival. Hildén (1964) noted this and also reported strong

Male displays of the red-breasted merganser, including (1) sprint, (2–3) salute-curtsy sequence on land, (4–7) salute-curtsy sequence in water (front view), and (8) curtsy (rear view).

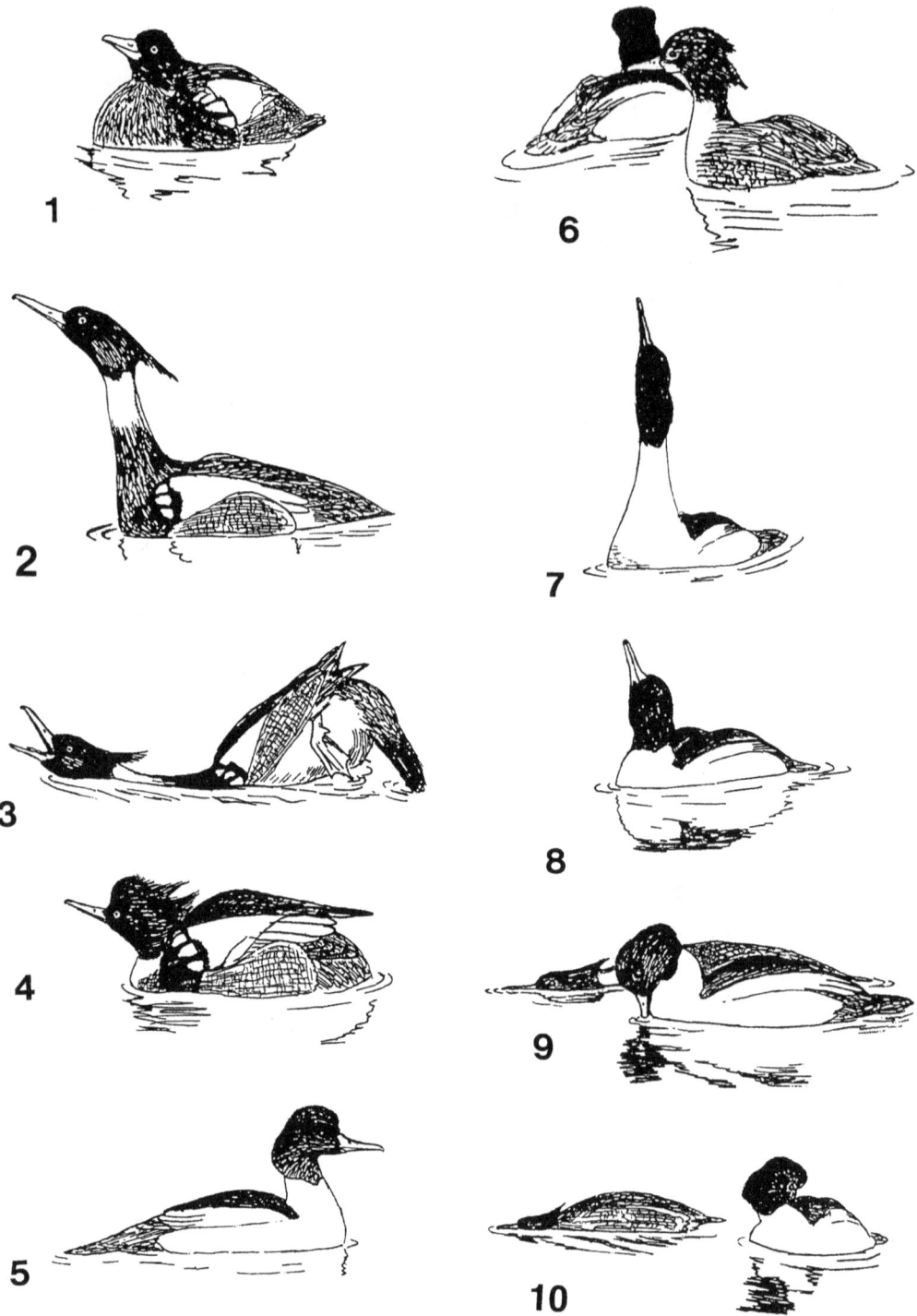

Male displays of the red-breasted (1–4) and common mergansers (5–10), including (1–4) salute-curtsy sequence (side view), (5) courtship call, (6) turning-of-the-back-of-the-head, (7) salute posture, (8–10) precopulatory sequence, including drinking, (9) water-twitching, and (10) preening-behind-the-wing.

brood-merging tendencies in this species. He noted broods of as many as 100 young attended by a single female. In some cases broods were also observed to be escorted, at least temporarily, by two females.

Postbreeding behavior. Males typically desert their mates early in incubation, and early observations of males apparently associated with broods have not been verified by more recent studies. Little information is available on postbreeding behavior and movements of red-breasted mergansers. There is no strong evidence that any substantial molt migration occurs, but very probably there is a general movement of males to brackish or saline waters prior to undergoing their flightless period. Hildén (1964) reported that, because of the species' sociality, small flocks were seen through the breeding season and it was difficult to determine the duration of the pair-bonds. Drakes began to flock when the hens began incubation, and by late June flocks of up to 30 were seen, including some presumably nonbreeding females. Most of these were gone by early July, suggesting a premolt migration by these birds.

Common Merganser
Mergus merganser Linnaeus 1758

Other vernacular names. Fish duck, goosander (*M. m. merganser*), sawbill

Range. Breeds in Iceland, central Europe, Scandinavia, Russia, and Siberia to Kamchatka and some of the Bering Sea islands, and in North America from southern Alaska and the southern Yukon eastward across central Canada to James Bay and across the Labrador Peninsula to New Brunswick, Nova Scotia, Prince Edward Island, and Newfoundland. Also southward in the western mountains to California, Utah, Arizona, New Mexico (rarely to Chihuahua, Mexico), Wyoming, Colorado, South Dakota (Black Hills), and northeastward to Minnesota, Wisconsin, Michigan, New York, Connecticut, Massachusetts, New Hampshire, and Maine. Winters both on salt and fresh water, from the Aleutian Islands to southern California, from Newfoundland to Florida, and in the interior wherever large rivers or deep lakes that are ice-free occur.

North American subspecies. *M. m. americanus* Cassin. American Merganser. Breeds in North America as indicated above.

 M. m. merganser (goosander) breeds in Europe and over much of Asia, is very slightly smaller, and has no known North American records.

Measurements (*M. m. americanus*). *Folded wing:* Palmer (1976): Males, ave. of 12, 278 mm, 269–285 mm; females, ave. of 12, 253 mm, 246–259 mm.

 Culmen: Palmer (1976): Males, ave. of 12, 55.8 mm, 54.5–59 mm; females, ave. of 12, 49.2 mm, 47–54 mm.

Weights (mass). Nelson and Martin (1953): 45 males ave. 3.5 lb. (1,588 g), max. 4.1 lb. (1,859 g); 29 females ave. 2.5 lb. (1,133 g), max. 3.9 lb. (1,769 g). Erskine (1971): 13 adult males (November) ave. 1,709 g (3.77 lb.); 13 adult females (October) ave. 1,220 g (2.69 lb.).

Identification

In the hand. Immediately recognizable as a merganser on the basis of its long, serrated bill, only the red-breasted merganser has a culmen length as long as this species, from 45 to 60 mm. However, the American merganser's bill differs in that it (1) has the nostril located in the middle third of the bill, (2) has feathering on the side of the lower mandible reaching nearly as far forward as that on the side of the upper mandible, (3) has a stouter and relatively shorter upper mandible that is usually no more than five times as long as high when measured from the mandible edge to the highest unfeathered area, and (4) has a larger, wider nail at its tip. Both sexes are substantially larger than the red-breasted merganser, with adult males and female

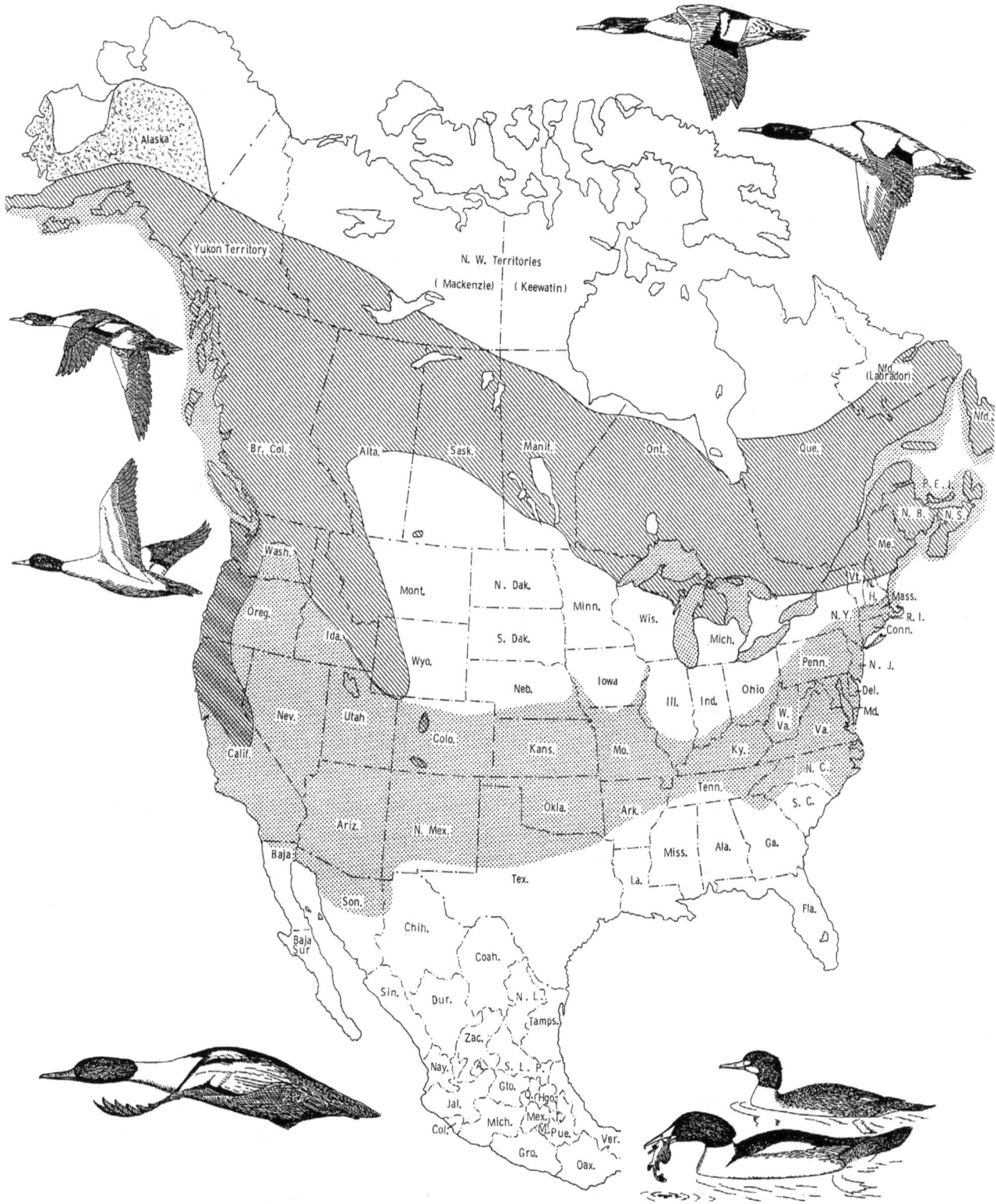

The North American breeding (hatched) and wintering (shaded) ranges of the common merganser. Regions of newly acquired or previously unrecognized breeding are shown by stippling.

common mergansers having minimum folded wing lengths of 270 and 236 mm, respectively, compared with respective maximum wing lengths of 260 and 230 mm in the red-breasted merganser.

In the field. When in nuptial plumage, a male common merganser is unmistakable, with its dark greenish head without a shaggy crest, its immaculate white (in spring) to pinkish-tinted underparts, and the absence of gray or black on its sides. Females and immature males appear to have generally grayish to white bodies, strongly contrasting with their reddish brown heads and necks. A clear white (rather than grayish) throat and a white line between the eye and the base of the bill can be seen at close range. In flight, a common merganser appears as a very large, long-necked streamlined duck with rapid wingbeats. It holds its head, neck, and body at the same level; both sexes exhibit a nearly pure white breast color and have almost entirely white underparts, including their underwing surface.

Age and Sex Criteria

Sex determination. Adult males have white middle and lesser wing-coverts and white tertials that are margined with black, rather than dark gray in these areas. First-year males reportedly also have the outer secondaries white and the inner ones gray (Bent, 1925), although more probably the reverse is true, with the outer gray secondaries conforming to the black outer secondaries of adult females. Erskine (1971) noted that not only are juvenile males larger than females but they also have a distinctive pale wing patch, formed by several outer secondary coverts.

Age determination. Adult males can be distinguished from first-year males by the latter's dark gray middle and lesser coverts and dark gray tertials. Adult females have solid gray, wider, and more rounded tertials and tertial coverts, while in immature females these feathers are narrower and have faded, wispy tips (Carney, 1964). Unlike females, males might become fertile toward the latter part of their first year of life, and thus a fully developed penis might not indicate a bird any older than one year. Immature birds have a well-developed bursa of Fabricius, which Erskine reported as absent in adults. Additional age and sex criteria have been reported by Anderson and Timken (1971), who noted that the juvenile rectrices are lost by the end of November and that adults and second-year birds have red bills and/or feet, whereas those of first-year birds are yellow to reddish orange.

Distribution and Habitat

Breeding distribution and habitat. The North American breeding distribution of the common merganser, like that of the red-breasted merganser, is transcontinental in character but is essentially confined to forested regions.

In Alaska, the common merganser breeds chiefly in the southern coastal area, occurring regularly as far west as Prince William Sound. Broods have also been reported on Kodiak Island. Very few definite

Common merganser, pair in breeding plumage

indications of breeding have been obtained for areas north of the Alaska Peninsula or in the interior, with the Yukon River and Paxon Lake seemingly approximating the northern limits of known breeding.

In Canada, breeding extends across the southern Yukon, southwestern Mackenzie District, and the wooded portions of British Columbia, the Prairie Provinces, and most of those parts of Ontario, Quebec, and Labrador south of the tree line, as well as in Newfoundland and the Maritime Provinces (Godfrey, 1986; Erskine, 1992). In British Columbia it is widespread, especially in coastal areas and the southern central interior, usually nesting below 750 meters in altitude (Davidson et al., 2008–2012).

South of Canada, breeding occurs in Washington (Yocom, 1951), Oregon (Gilligan et al., 1994), and California (Foreman, 1976). There is also an extension of the species' breeding range southward along the Rocky Mountains through Idaho, Montana, and Wyoming, with widespread breeding in western Colorado (Kingery, 1998), and local breeding in South Dakota's Black Hills. It has also bred in Arizona and in Chihuahua, Mexico (Brown, 1990). East of the Great Plains, it also nests regularly in northern Minnesota (Lee et al., 1964a), rarely in Wisconsin (Jahn and Hunt, 1964), commonly in northern and central Michigan (Brewer, McPeek, and Adams, 1991), and in New England north to southern Maine (Palmer, 1949). South of New England there have been a few isolated breeding records, such as in western New York, New Hampshire, Massachusetts, Connecticut, and Virginia. There are some historic breeding records in the southeastern states (Kiff, 1998).

The preferred breeding habitat of this species consists of ponds associated with upper portions of rivers in forested regions and of clear freshwater lakes with forested shorelines. Clear water is needed for visual foraging by both adult and young birds. Hildén (1964) reported that this species is an inland rather than marine nester, breeding along extensive waters with barren shores and rivers with clear water, almost entirely within the forest zone. It is not socially attracted to gulls, and its nesting distribution seems primarily determined by nest site availability and landscape characteristics. Islands are favored breeding areas, especially if they are rather barren and boulder-covered, the boulders providing alternate nest sites if tree cavities are unavailable.

Wintering distribution and habitat. In contrast to the red-breasted merganser, this species preferentially seeks out fresh water during winter. It is found uncommonly in Alaska as far west as the Aleutians (Murie, 1959; Gibson and Byrd, 2007), and along coastal southeastern Alaska, where it ranges from being common to plentiful (Gabrielson and Lincoln, 1959).

In Canada, it winters commonly along coastal British Columbia, with small numbers found in the interior of that province, in Alberta, on the Great Lakes, in southern Quebec, and coastally along Newfoundland and the Maritime Provinces (Godfrey, 1986).

South of Canada, varying numbers of common mergansers can be found in winter on almost any large lake, reservoir, or river that remains partly ice-free all winter, southward as far as southern California, Texas, and the Carolinas. In the Chesapeake Bay region, they are common but locally distributed, with most of them occurring on fresh estuarine bays or bay marshes and a few ranging into slightly brackish estuarine bays or river marshes (Stewart, 1962). Very few common mergansers winter as far south as Mexico, and most of the records are from its northern parts (Leopold, 1959).

The most characteristic type of winter habitat consists of the mouths and the upper estuarine regions of rivers (Dementiev and Gladkov, 1967). Like red-breasted mergansers, this species needs relatively transparent waters for efficient foraging, and it congregates wherever fish are to be found in good numbers.

General Biology

Age at maturity. Maturity is probably attained the second year of life (Dementiev and Gladkov, 1967). Erskine (1971) reported that at least six females were found to breed at two years of age, but none were known to breed their first year, although males might be sexually mature by that age.

Pair-bond pattern. Pair-bonds are reestablished each winter, starting about November or December. The time of pair-bond breakup is difficult to determine, but Hildén (1964) reported seeing pairs as late as early June. He did not observe any cases of males remaining with their mates after the young had hatched, although some early observations have suggested that this might sometimes occur. Instead, it appears that the male usually leaves the nesting area about the time the female begins to incubate (White, 1957).

Common merganser, pair in flight

Nest location. Hildén's (1964) summary of 113 nest sites is probably representative, although trees large enough to support cavity nests were generally lacking from his study area. Of the total nests found, 68 percent were beneath boulders, 18 percent were in buildings, 13 percent were under dense, mat-like junipers, and 1 percent were under *Hippophae* bushes. The primary requirement appears to be concealment from above and associated darkness in the nest cavity. Common mergansers are apparently not attracted to gull colonies for nesting, and on large islands they tend to nest near shore on headlands. Islets located where waters become free of ice early in the spring season are also favored, according to Hildén.

In Iceland, Bengtson (1970) noted that six of ten nests found were in holes, two were under high shrubs, and two were in other cover. Where nests are located in trees, a variety of species have been used, including oak, beech, chestnut, sycamore, basswood, willow, and alder (Bauer and Glutz, 1969). If artificial nesting boxes are used, their preferred dimensions are 23 to 28 cm wide and 85 to 100 cm high. The entrance should measure 12 × 12 cm, and be located 50 to 60 centimeters above the base of the box (Grenquist, 1953). Common mergansers are apparently not influenced by gull colonies while nesting, and on large islands they tend to nest near shore on headlands. Islets located where waters become free of ice early in the spring season are also favored, according to Hildén.

Clutch size. Hildén (1964) reported an average clutch size of 9.37 eggs for 35 clutches, with an observed range of 6 to 12 eggs and a modal clutch of 9 eggs. Von Hartmann (in Bauer and Glutz, 1969) reported a mean clutch size of 9.2 eggs for 104 Finnish clutches. The eggs are deposited daily (Bauer and Glutz, 1969). Intraspecific brood parasitism is common, with clutches of up to 19 eggs reported.

Incubation period. The incubation period is estimated at 32 to 35 days (based on various sources in Bauer and Glutz, 1969), 30 to 32 days (Kear, 2005), and 28 to 35 days (Mallory and Metz, 1999).

Fledging period. From 60 to 70 days are required to attain fledging (Dementiev and Gladkov, 1967; Kear, 2005); the period is also estimated at 60 to 75 days (Erskine, 1972). These are among the longest fledging periods of sea ducks.

Nest and egg losses. Hildén (1964) reported that during a two-year period of study, 86 percent of 29 nests hatched. When records of other nests studied were added, a total nesting success of 84 percent was determined for 73 nests. Considering total eggs, a hatching success rate of 77 percent was estimated for this species. Crows and ravens were responsible for a high rate of egg losses prior to the onset of incubation, in spite of the fine concealment of most nests. This was attributed to the presence of conspicuous white down near the nest entrance and to the apparent memory that these intelligent corvids have of nest-site locations that are used by mergansers year after year.

Juvenile mortality. Relatively little information on prefledging losses of this species is available. Hildén (1964) noted that shortly after hatching their young, female common mergansers and their broods left his study area for unknown reasons. Some returned when the young were at least half grown, but the high mobility of this species' broods makes estimates of their numbers in any area very difficult. Foreman (1976) estimated a duckling mortality rate of 20 to 50 percent, and Wood (1986) estimated that 63 to 83 percent of the birds hatched were reared. Average brood size has generally ranged from 8 to 11 in various studies, suggesting a fairly high survival rate of the young, although brood mergers make such conclusions uncertain.

Adult mortality. An early estimate of annual adult mortality was that of Boyd (1962), who calculated a 40 percent annual mortality rate (60 percent survival) for birds wintering in Britain, and a mean life expectancy of 2 years. Later recoveries of 250 banded birds have suggested a mean survival duration of 2.26 years. This might be an underestimate, since the longevity record for wild males is 12 years, 6 months, and for females 13 years, 10 months; some wild females have been known to still be breeding at 6 years of age (Mallory and Metz, 1990).

Common merganser, female

General Ecology

Food and foraging. The controversies and emotions generated by the fish-eating tendencies of this species are considerable, and a judicious choice of references can allow the writer to cast the common merganser in almost any role that might be desired. Perhaps the fairest method is to consider the evidence on a region-by-region basis, since major regional and habitat differences in available food sources are obviously present.

In Alaska, British Columbia, and adjacent Washington, the primary concern has been the influence of common mergansers on the salmon and trout fisheries. Relatively few samples are available from Alaskan waters, but Fritsch and Buss (1958) examined 55 birds from Unakwik Inlet. Unidentified fish remains made up the largest single volumetric amount of foods, but of identified food materials various sculpins (Cottidae) made up the greatest volume (69 cubic cm), with the great sculpin (*Myoxocephalus*) adding another 61 cubic cm, shrimp totaling 54.5 cubic cm, and blennies (*Anoplorchus*) 14 cubic cm. Salmon eggs were present in trace amounts in seven birds, and salmon fry were present in similar quantities in three birds.

In British Columbia, Munro and Clemens (1937) examined the food taken by 363 common mergansers and found that in order of relative importance it consisted of freshwater sculpins, salmon eggs, salmonid fish (char, trout, salmon), sticklebacks, freshwater coarse fish, and various marine fish. These authors concluded that in British Columbia the common merganser did exert a significantly detrimental effect on salmon. Studies in Washington, as summarized by Meigs and Rieck (1967), indicate that local damage to trout fisheries can occur, particularly on trout-planted waters. These authors found that a juvenile bird consumed an average of 0.77 lb. of fish per day for 83 days, similar to an estimate by White (1937) that a young merganser daily consumes the equivalent of a third of its weight, and a comparable estimate of 32 to 40 percent by Feltham (1995).

More recent studies by White (1957) and by Latta and Sharkey (1966) suggest that food equal to about 20 to 28 percent of the body weight is consumed each day in older mergansers, but their birds did not maintain their original weight during the study periods. Studies by White (1957) in the Maritime Provinces of Canada indicate a rather high depredation by the common merganser on salmon streams. Among samples of 724 common mergansers, salmon remains accounted for 5 to 91 percent of the fish remains, and occurred in 45 to 96 percent of the stomachs examined. White estimated that a single merganser might consume 72 pounds of fish before attaining its full growth.

Trout rivers in Michigan are sometimes utilized heavily by mergansers, according to Salyer and Lagler (1940), who examined 315 specimens from various parts of Michigan. They found that, on trout streams, trout predominated in the merganser stomach samples, and judged that trout were preferentially selected among other available organisms in the streams. However, samples from nontrout waters indicated that in such areas the mergansers were innocuous and at times beneficial.

Mergansers collected in Minnesota, Nebraska, and South Dakota were examined by Timken and Anderson (1969). Among 222 birds, only 19 of 151 identified fish remains represented game species, and the most important food was gizzard shad (*Dorosoma*), which composed 37 percent of the total food weight. Freshwater drum (*Aplodinotus*) and white bass (*Roccus*) were sequentially next in importance; these three species made up about 60 percent of the total number of fish found.

A similar finding was provided by Alcorn (1953), based on a sample of 110 stomachs from Nevada. Of a total of 267 fish present, 76 percent were various rough fish, mostly consisting of carp (*Cyprinus*). Heard and Curd (1959) likewise reported that 80 percent of the fish found in mergansers obtained from Lake Carl Blackwell in Oklahoma were various rough or forage fishes. Huntington and Roberts (1959) found no evidence that the common merganser was a menace to sport fishing in New Mexico and positively correlated the amounts of various fish eaten with their availability as indicated by fisheries studies.

The maximum sizes of fish taken by mergansers are rather astonishing. Alcorn (1953) reported finding carp up to 12.75 inches long, Salyer and Lagler (1940) noted a case of a merganser with a 14-inch, 15-ounce brown trout, and Coldwell (1939) reported a 22-inch eel eaten by a merganser. Wick and Rogers (1957) described a female merganser that had choked to death on a sculpin measuring 14.9 centimeters and weighing 64.3 grams. Latta and Sharkey (1966) judged that girth (up to about 200 mm) rather than length probably determined the maximum size of fish that could be swallowed. Captive birds seemed to prefer small trout

Common merganser, male

to larger ones, and when given a choice of trout, sculpins, and creek chubs (*Semotilis*), they consumed all three species, with the sculpins in somewhat smaller numbers.

Mergansers catch their prey visually, and in clear water can see fish up to 10 feet away (White, 1937). They prefer to feed in fairly shallow waters from 1.5 to 6 feet deep. When diving for food they generally remain under water for 10 to 20 seconds but occasionally remain submerged up to 45 seconds (Salyer and Lagler, 1940). They have been reported to dive as deep as 30 feet (Heard and Curd, 1959). White (1957) described cooperative foraging by flocks of 20 or more mergansers, which would form a long line parallel to the shore of a river or shallow lake. With much wing-splashing the flock would advance, and then suddenly the birds would dive and catch the fish that had been thus concentrated. Single flocks of as many as 70 birds were observed foraging by this method.

Mallory and Metz (1999) summarized the results of food consumption studies from three interior and three coastal sites. At the two coastal sites, the three most important foods (by descending volume or numbers of prey present) were (1) by prey item numbers present, Nova Scotia: Atlantic salmon, blacknose dace, and white sucker, and (2) by numbers present, coastal British Columbia: sculpin, stickleback, and salmonids. At the four interior sites, the three most important foods were (1) by numbers present, British Columbia: sculpin, chub, and lake shiner, (2) by numbers present, New Mexico: threadfin shad, gizzard shad, and unidentified shad, (3) by volume, Michigan: brown trout, rainbow trout and, white sucker, and (4) by volume, Oklahoma: gizzard shad, freshwater drum, and white crappie.

The general conclusion from most of these studies is that the common merganser is an opportunistic forager, feeding on such species as are fairly common, of a suitable size, and readily captured. In most cases these consist of rough fish rather than game fish, but in areas specifically managed for trout or salmon production, mergansers will probably concentrate on this easily available supply of food.

The only detailed study on duckling foods is that of White (1957), who analyzed the stomach contents of 118 ducklings. These included nearly 1,400 insects, of which more than 93 percent were mayflies (Ephemeroptera). There were also more than 300 fish present, 70 percent of which were species other than salmon or trout. A transition from invertebrate foods to fish begins at about 12 days of age.

Sociality, densities, territoriality. Probably most mergansers breed in relatively well-isolated and well-separated situations, since an adequate supply of food for the female and developing young is required. Parmelee (1954) noted that 1 to 2 pairs along a 16-mile stretch of the Sturgeon River in Michigan seemed typical, although on some Lake Michigan islands a more concentrated nesting of this species, and of the red-breasted merganser, occurs. There is no evidence of territorial behavior, and there are cases of up to three females nesting in the same tree (Cramp and Simmons, 1977).

A comparable density situation was described by Hildén (1964) in the Valassaaret group of islands in the Gulf of Bothnia, Finland. In 1962 this island group had an estimated breeding population of 34 pairs scattered over the 6 square kilometers of land area, or nearly 6 pairs per square kilometer. However, many of the tiny rocky islets supported nesting pairs, and thus nest site availability as well as proximity to suitable foraging areas were probably important factors determining distribution and density characteristics.

Interspecific relationships. When both species are on fresh water, common and red-breasted mergansers have similar foraging tendencies and consume nearly identical foods. However, for most of the year these species are well separated ecologically from one another, and it is doubtful that much competition occurs. Double-crested cormorants also are freshwater and marine fish-eaters, but are rather rarely found in association with nesting common mergansers. Egg predators such as crows and ravens, and duckling predators such as the larger gulls and jaegers, no doubt account for substantial mortality to eggs and broods; at least this is indicated by such studies as have been done (Hildén, 1964). Merganser ducklings seem more agile than those of most species in eluding predators, and furthermore appear to be less sensitive to chilling effects of severe weather. Eagles, owls, minks, and loons were mentioned by White (1957) as possible enemies of merganser ducklings.

General activity patterns and movements. Common mergansers are well known to be daytime foragers. Timken and Anderson (1969) reported that fall migrants in South Dakota seem to confine their feeding to morning hours, while during winter and spring they forage in the early morning and again in the late afternoon. A similar morning and afternoon foraging periodicity was noted by Salyer and Lagler (1940) in Michigan. According to White (1957), the most active period of feeding is just before twilight, and there is usually a resting period of at least two hours at midday. Nilsson (1966) mentioned that this species spent less time foraging than did common goldeneyes in the same locality.

Social and Sexual Behavior

Flocking behavior. During fall, the size of migrant flocks is usually rather small. Timken and Anderson (1969) indicated that groups of 8–9 birds were typical, and the groups never exceeded 30. Salyer and Lagler (1940) also mentioned that foraging is usually done in small flocks of fewer than 12 birds, with such groups often having 2–3 adult males and the rest females or female-like immatures. These flocks do not appear to feed cooperatively, but probably the success of birds feeding in small groups is greater than that of single birds, since these are seen infrequently. Nilsson (1966) mentioned seeing aggression among feeding flocks, and in one case observed a bird stealing a fish from another. As spring approaches, flock sizes further decrease, and many birds are by then paired.

Pair-forming behavior. Pair-forming displays can be seen in wintering areas and also among spring migrants. It is marked by a great deal of surface chasing among the males, somewhat resembling the "sprints" of the red-breasted merganser. The most common male courtship call is a guitar-like note, uttered with the neck partly stretched, the throat slightly enlarged, and the head feathers fluffed. A second call, a bell-like note, is uttered during a sudden vertical stretching of the head and neck in a "salute" posture. Males also at times suddenly kick a jet of water backward some distance, but there is no associated head movement. During aquatic courtship the males produce a rather faint *uig-a* sound also reminiscent of the twanging of a guitar string, and sometimes the females utter harsh *karrr* notes.

The male often swims ahead of the female, with his tail cocked diagonally or flat on the water, and turns the back of his head toward the female, especially if she is inciting. The inciting behavior of this species is much like that of the red-breasted merganser. It consists of a loud, harsh call, repeated once or twice, and is associated with rapid forward swimming as each note is uttered (Johnsgard, 1965). Short display flights have been seen on a few occasions (Johnsgard, 1955); these terminate in a long, "skidding" stop near the courted females.

Copulatory behavior. Copulation in this species is normally preceded by a mutual drinking display. The female then assumes a prone posture, after which the male performs a lengthy series of drinking, preening, shaking, and similar "comfort movements" that differ little if at all from their nondisplay counterparts

(Johnsgard, 1965). Nilsson (1966) also noted that the male's precopulatory behavior was seemingly unritualized and involved preening, drinking, and bill-dipping movements. Mounting is not preceded by any obvious displays, and, after treading, either the male immediately releases the female (Johnsgard, 1965) or the pair rotates in the water (Nilsson, 1966). In my observations, the male then swims away from the female, while uttering his courtship call repeatedly and keeping the back of his head oriented toward her. Nilsson observed only bathing after copulation.

Nesting and brooding behavior. Females remain fairly gregarious during the early stages of the nesting season, while they are searching for suitable nest sites, and in areas providing numerous suitable sites assemblages of nesting females might occur. The males might remain in the general vicinity of the nest during incubation and are sometimes also seen in the presence of broods, but they do not defend the young (Bauer and Glutz, 1969). Often the males leave their mates and drift downstream, leaving the female and brood to forage in the upper reaches of the river (White, 1957). While incubating, females usually leave their nests for a short time each day, often between 7:00 a.m. and 2:00 p.m., for periods of 15 to 90 minutes.

After hatching occurs, the female typically remains in the nest for 1.5 to 2 days before leading the young to water. Several undocumented reports of females carrying young to water have been described, and younger ducklings can often be seen riding on the back of a swimming female. The young are highly precocial, and broods are very mobile, a situation enhanced by the tendency of the female to carry part of her brood on her back. While still fairly young the ducklings begin increasingly to shift for themselves and seem to survive fairly well without direct parental attention. The female often deserts her brood to begin the postnuptial molt before the young have fledged. At this time the ducklings often begin to gather into larger assemblages (Bauer and Glutz, 1969). An important facet of the habitat for flightless young is the presence of resting or roosting places closely adjacent to water at least 2 feet deep, where the birds can rapidly escape from danger (White, 1957).

Postbreeding behavior. Molt migrations have not been well documented for this species. However, White (1957) noted that most yearling females apparently leave their breeding streams before molting, and that both adult and immature males apparently move out to sea to complete their molts. In Keith County, Nebraska, several hundred miles from the nearest Colorado or Wyoming breeding areas, 10 to 15 mostly adult common mergansers annually appear below Kingsley Dam in June, and undergo their flightless period below the dam on the outflow Lake Ogallala, where they feed on dead and injured fish that pass through the Kinsley Dam spillway or turbine.

During the fall, aggregations of fairly large numbers of birds occur on favored foraging areas such as clearwater rivers and lakes, and a leisurely movement southward begins. The birds gradually move to large lakes and reservoirs or other ice-free waters to spend the winter.

III. References

Common merganser, pair in flight

General and Multispecies Sea Duck References

Species Monographs in *The Birds of North America*

Bordage, D., and J-P. L. Savard. 1995. Black Scoter. In *The Birds of North America*, No. 177. (A. Poole and F. Gill, eds.). Philadelphia, PA: The Academy of Natural Sciences, and Washington, DC: American Ornithologists' Union. 20 pp. (Includes about 130 citations.)

Brown, P. W., and L. H. Fredrickson. 1997. White-winged Scoter. In *The Birds of North America*, No. 274. (A. Poole and F. Gill, eds.). Philadelphia, PA: The Academy of Natural Sciences, and Washington, DC: American Ornithologists' Union. 28 pp. (Includes about 160 citations.)

Chilton, G. 1997. Labrador Duck. In *The Birds of North America*, No. 307. (A. Poole and F. Gill, eds.). Philadelphia, PA: The Academy of Natural Sciences, and Washington, DC: American Ornithologists' Union. 12 pp. (Includes about 50 citations.)

Dugger, B. D., K. M. Dugger, and L. H. Fredrickson. 1994. Hooded Merganser. In *The Birds of North America*, No. 98. (A. Poole and F. Gill, eds.). Philadelphia, PA: The Academy of Natural Sciences, and Washington, DC: American Ornithologists' Union. 24 pp. (Includes about 100 citations.)

Eadie, J. M., J-P. L. Savard, and M. L. Mallory. 2000. Barrow's Goldeneye. In *The Birds of North America*, No. 548. (A. Poole and F. Gill, eds.). Philadelphia, PA: The Academy of Natural Sciences, and Washington, DC: American Ornithologists' Union. 32 pp. (Includes about 120 citations.)

Eadie, J. M., M. L. Mallory, and H. G. Lumsden. 1995. Common Goldeneye. In *The Birds of North America*, No. 170. (A. Poole and F. Gill, eds.). Philadelphia, PA: The Academy of Natural Sciences, and Washington, DC: American Ornithologists' Union. 32 pp. (Includes about 150 citations.)

Fredrickson, L. H. 2001. Steller's Eider. In *The Birds of North America*, No. 571. (A. Poole and F. Gill, eds.). Philadelphia, PA: The Academy of Natural Sciences, and Washington, DC: American Ornithologists' Union. 24 pp. (Includes about 125 citations.)

Gauthier, G. 1993. Bufflehead. In *The Birds of North America*, No. 67. (A. Poole and F. Gill, eds.). Philadelphia, PA: The Academy of Natural Sciences, and Washington, DC: American Ornithologists' Union. 24 pp. (Includes about 100 citations.)

Goudie, R. I., G. J. Robertson, and A. Reed. 2000. Common Eider. In *The Birds of North America*, No. 546. (A. Poole and F. Gill, eds.). Philadelphia, PA: The Academy of Natural Sciences, and Washington, DC: American Ornithologists' Union. 32 pp. (Includes about 225 citations.)

Mallory, M., and K. Metz. 1999. Common Merganser. In *The Birds of North America*, No. 442. (A. Poole and F. Gill, eds.). Philadelphia, PA: The Academy of Natural Sciences, and Washington, DC: American Ornithologists' Union. 28 pp. (Includes about 175 citations.)

Petersen, M. R., J. B. Grand, and C. P. Dau. 2000. Spectacled Eider. In *The Birds of North America*, No. 547. (A. Poole and F. Gill, eds.). Philadelphia, PA: The Academy of Natural Sciences, and Washington, DC: American Ornithologists' Union. 24 pp. (Includes about 100 citations.)

Robertson, G. J., and J-P. L. Savard. 2002. Long-tailed Duck. In *The Birds of North America*, No. 651. (A. Poole and F. Gill, eds.). Philadelphia, PA: The Academy of Natural Sciences, and Washington, DC: American Ornithologists' Union. (Includes about 200 citations.)

Robertson, G. J., and R. I. Goudie. 1999. Harlequin Duck. In *The Birds of North America*, No. 466. (A. Poole and F. Gill, eds.). Philadelphia, PA: The Academy of Natural Sciences, and Washington, DC: American Ornithologists' Union. (Includes about 225 citations.)

Savard, J.-P. L., D. Bordage, and A. Reed. 1998. Surf Scoter. In *The Birds of North America*, No. 363. (A. Poole and F. Gill, eds.). Philadelphia, PA: The Academy of Natural Sciences, and Washington, DC: American Ornithologists' Union. 28 pp. (Includes about 200 citations.)

Suydam, R. S. 2000. King Eider. In *The Birds of North America*, No. 491. (A. Poole and F. Gill, eds.). Philadelphia, PA: The Academy of Natural Sciences, and Washington, DC: American Ornithologists' Union. 28 pp. (Includes about 200 citations.)

Titman R. D. 1999. Red-breasted Merganser. In *The Birds of North America*, No. 443. (A. Poole and F. Gill, eds.). Philadelphia, PA: The Academy of Natural Sciences, and Washington, DC: American Ornithologists' Union. 24 pp. (Includes about 150 citations.)

Taxonomic Works and General Surveys

American Ornithologists' Union. 1998. *Check-list of North American Birds.* 7th edition. American Ornithologists' Union, Washington, DC.

Audubon, J. J. 1883. *Ornithological Biography.* Volume 4. Adam and Charles Black, Edinburgh, Scotland.

Baldassarre, G. A, 2014. *Ducks, Geese, and Swans of North America.* Rev. ed. Johns Hopkins University Press. Baltimore, MD.

Bauer, K. M., and U. N. Glutz von Blotztheim. 1968. *Handbuch der Vogel Mitteleuropas.* Band 2. Frankfurt am Main: Akademische Verlagsgesellschaft.

Bauer, K. M., and U. N. Glutz von Blotztheim. 1969. *Handbuch der Vogel Mitteleuropas.* Band 3. Frankfurt am Main: Akademische Verlagsgesellschaft.

Bellrose, F. 1976. *The Ducks, Geese, and Swans of North America.* 2nd ed. Wildlife Management Institute, Washington, DC.

Bellrose, F. 1980. *The Ducks, Geese, and Swans of North America.* 3rd ed. Wildlife Management Institute, Washington, DC.

Bent, A. C. 1925. *Life Histories of North American Wild Fowl.* Part 2. US National Museum Bulletin 130. US Government Printing Office, Washington, DC.

Brooks, T. 2000. Extinct species. Pp. 701–708 in *Threatened Birds of the World.* Lynx Edicions and BirdLife International, Barcelona and Cambridge, UK.

Carney, S. M. 1964. *Preliminary Key to Age and Sex Identification by Means of Wing Plumage.* US Fish and Wildlife Service Special Scientific Report: Wildlife No. 82.

Cramp, S., and K. E. L. Simmons. 1977. *Handbook of the Birds of Europe, the Middle East, and North Africa: The Birds of the Western Palearctic, Volume 1, Ostrich to Ducks.* Oxford University Press, Oxford, UK.

Delacour, J. 1959. *The Waterfowl of the World.* Volume 3. Country Life, London.

Delacour, J., and E. Mayr. 1945. The family Anatidae. *Wilson Bulletin* 57:3–55.

del Hoyo, J., A. Elliott, and J. Sargatal. 1992. *Handbook of the Birds of the World. Vol. 1. Ostrich to Ducks,* Barcelona: Lynx Edicions.

Gillham, E., and B. Gillham. 1996. *Hybrid Ducks: A Contribution towards an Inventory.* Hythe Printers, Hythe, Kent, England, UK.

Gray, A. P. 1958. *Bird Hybrids: A Check-list with Bibliography.* Technical Communication 13. Commonwealth Agricultural Bureau, Farnham Royal, Bucks, England, UK.

Humphrey, P. S. 1958. Classification and systematic position of the eiders. *Condor* 60: 129–135.

Johnsgard, P. A. 1960a. Hybridization in the Anatidae and its taxonomic implications. *Condor* 62: 25–33. http://digitalcommons.unl.edu/biosciornithology/71

Johnsgard, P. A. 1960b. Classification and evolutionary relationships of the sea ducks. *Condor* 62: 426–433. http://digitalcommons.unl.edu/biosciornithology/70

Johnsgard, P. A. 1961a. Tracheal anatomy of the Anatidae and its taxonomic significance. *Wildfowl Trust Annual Report* 12: 58–69.

Johnsgard, P. A. 1961b. The taxonomy of the Anatidae–A behavioural analysis. *Ibis* 103a: 71–85. http://digitalcommons.unl.edu/johnsgard/29

Johnsgard, P. A. 1963. Behavioral isolating mechanisms in the family Anatidae. Pp. 531–543. *Proc. XIII International Ornithological Congress.* http://digitalcommons.unl.edu/johnsgard/23

Johnsgard, P. A. 1965. *Handbook of Waterfowl Behavior.* Cornell University Press, Ithaca, NY. http://digitalcommons.unl.edu/bioscihandwaterfowl/7/

Johnsgard, P. A. 1975. *Waterfowl of North America.* Indiana University Press, Bloomington, IN. http://digitalcommons.unl.edu/biosciwaterfowlna/1/

Johnsgard, P. A. 1978. *Ducks, Geese, and Swans of the World.* University of Nebraska Press, Lincoln. http://digitalcommons.unl.edu/biosciducksgeeseswans/

Johnsgard, P. A. 1979. *Anseriformes* section (Anatidae and Anhimidae). Pp. 425–506, in *Check-list of the Birds of the* World (E. Mayr, ed.). Harvard Univ. Press, Cambridge, MA. http://digitalcommons.unl.edu/johnsgard/32

Johnsgard, P. A. 1997. *The Avian Brood Parasites: Deception at the Nest.* Oxford University Press, New York. 409 pp.

Johnsgard, P. A. 2010. *Ducks, Geese, and Swans of the World.* Rev. ed., with a supplement: "The World's Waterfowl in the 21st Century." University of Nebraska–Lincoln DigitalCommons and Zea Books. 498 pp. http://digitalcommons.unl.edu/biosciducksgeeseswans/)

Kear, J., ed. 2005. *Ducks, Geese, and Swans. Species Accounts Cairina–Mergus.* Oxford University Press, Oxford, England, UK Vol. 2.

Kessler, L. G., and J. C. Avise. 1984. Systematic relationships among waterfowl (Anatidae) inferred from restriction endonuclease analysis of mitochondrial DNA differentiation in selected avian and other vertebrate genera. *Systematic Zoology* 33: 370–380.

Kortright, F. H. 1942. *The Ducks, Geese, and Swans of North America.* Wildlife Management Institute, Washington, DC.

Livezey, B. C. 1995. Phylogeny and evolutionary ecology of modern seaducks (Anatidae: Mergini). *Condor* 97: 233–255.

Livezey, B. C. 1997. A phylogenetic classification of waterfowl (Aves: Anseriformes), including selected fossil species. *Annals of the Carnegie Museum* 66: 457–496.

Madge, S., and H. Burn. 1988. *Waterfowl: An Identification Guide to the Ducks, Geese, and Swans of the World.* Houghton Mifflin, Boston, MA.

Mosby, H. S. (ed.). 1967. *Wildlife Investigational Techniques.* 2nd ed. Wildlife Society, Washington, DC.

Myres, M. T. 1959. The behaviour of the sea-ducks and its value in the systematics of the tribes Mergini and Somateriini of the family Anatidae. PhD dissertation, University of British Columbia, Vancouver.

Palmer, R. S. (ed.). 1976. *Handbook of North American Birds*, Vol. 3: Waterfowl, Part 2. Yale University Press, New Haven, CN.

Phillips, J. C. 1926. *A Natural History of the Ducks.* Vol. 4. Houghton Mifflin, Boston, MA.

Todd, F. S. 1996. *Natural History of the Waterfowl.* Ibis Publishing, Vista, CA.

Regional Bird Surveys and Faunal Lists

Angell, T., and K. C. Balcomb, III. 1982. *Marine Birds and Mammals of Puget Sound.* University of Washington Press, Seattle.

Bailey, A. M. 1948. *Birds of Arctic Alaska.* Popular Series 8. Colorado Museum of Natural History, Denver.

Bannerman, D. A. 1958. *Birds of the British Isles*, Vol. 7. Oliver and Boyd, Edinburgh and London.

Bergman, R. D., R. L. Howard, K. F. Abraham, and M. W. Weller. 1977. *Water Birds and Their Wetland Resources in Relation to Oil Development at Storkersen Point, Alaska.* Resource Publication 129. US Fish and Wildlife Service, Washington, DC.

Boertmann, D. 1994. *An Annotated Checklist to the Birds of Greenland.* Meddelelser om Grønland, Bioscience 38. Commission for Scientific Research in Greenland, Copenhagen, Denmark.

Brewer, R., G. A. McPeek, and R. J. Adams, Jr. 1991. *The Atlas of Breeding Birds of Michigan.* Michigan State University Press, East Lansing.

Buchanan, J. B. 2006. *Nearshore Birds in Puget Sound*. Technical Report 2006–05. Puget Sound Nearshore Partnership, Olympia, WA.

Butler, R. W., and R. W. Campbell. 1987. *The Birds of the Fraser River Delta: Populations, Ecology, and International Significance*. Occasional Paper 65. Canadian Wildlife Service, Ottawa.

Butler, R. W., N. K. Dawe, and D. E. C. Trethewey. 1989. The birds of estuaries and beaches in the Strait of Georgia. Pp. 26–34 in K. Vermeer and R. W. Butler, eds. *The Ecology and Status of Marine and Shoreline Birds in the Strait of Georgia, British Columbia: Proceedings of a Symposium*. Pacific Northwest Bird and Mammal Society and the Canadian Wildlife Service, Sidney, BC.

Byrd, G. V. 1992. *Status of Sea Ducks in the Aleutian Islands, Alaska*. US Fish and Wildlife Service, Adak, AK.

Byrd, G. V., D. L. Johnson, and D. D. Gibson. 1974. The birds of Adak Island, Alaska. *Condor* 76: 288–300.

Cadman, M. D., P. J. F. Engels, and F. M. Helleiner, eds. 2016. *Atlas of the Breeding Birds of Ontario*. 2nd ed. Federation of Ontario Naturalists and Long Point Bird Observatory, Bird Studies Canada, Environment Canada, and Ontario Field Ornithologists, Ontario Ministry of Natural Resources, and Ontario Nature, Toronto.

Campbell, R. W., N. K. Dawe, I. McTaggart-Cowan, J. M. Cooper, G. W. Kaiser, and M. C. E. McNall. 1990. *The Birds of British Columbia. Volume 1: Nonpasserines, Introduction and Loons through Waterfowl*. University of British Columbia Press, Vancouver.

Canadian Wildlife Service Waterfowl Committee. 2013. *Population Status of Migratory Game Birds in Canada*. Canadian Wildlife Service Migratory Birds Regulatory Report No. 40. Canadian Wildlife Service, Ottawa. http://www.ec.gc.ca/rcom-mbhr/default.asp?lang=En&n=B2A654BC-1

Conant, B., and D. J. Groves. 2002. *Alaska–Yukon Waterfowl Breeding Population Survey, May 17 to June 9, 2002*. US Fish and Wildlife Service, Juneau, AK.

Conant, B., D. J. Groves, and T. J. Moser. 2007. *Distribution and Abundance of Wildlife from Fixed-wing Aircraft Surveys in Nunavut, Canada, June 2006*. US Fish and Wildlife Service, Juneau, AK.

Conant, B., F. Roetker, and D. J. Groves. 2006. *Distribution and Abundance of Wildlife from Fixed-wing Aircraft Surveys on Victoria Island and Kent Peninsula, Nunavut, Canada, June 2005*. US Fish and Wildlife Service, Juneau, AK.

Conover, H. B. 1926. Game birds of the Hooper Bay region of Alaska. *Auk* 43: 162–180.

Cornish, B. J., and D. L. Dickson. 1996. *Distribution and Abundance of Birds on Western Victoria Island, 1992 to 1994*. Technical Report Series 253. Canadian Wildlife Service, Prairie and Northern Region, Edmonton, AB.

Dau, C. P., and E. J. Mallek. 2009. *Aerial Survey of Emperor Geese and Other Waterbirds in Southeastern Alaska, Spring 2009*. US Fish and Wildlife Service, Anchorage, AK.

Dau, C. P., and T. A. Schafer. 1996. *Aerial Duck Population Survey on the Izembek and Pavlof Units of the Izembek National Wildlife Refuge Complex, 1996*. US Fish and Wildlife Service, Cold Bay, AK.

Davidson, P. J. A., R. J. Cannings, A. R. Couturier, D. Lepage, and C. M. D. Corrado, eds. 2008 et seq. *The Atlas of the Breeding Birds of British Columbia*. Bird Studies Canada. Delta, BC. Species accounts are available at http://www.birdatlas.bc.ca/accounts

Dementiev, G. P., and N. A. Gladkov, eds. 1967. *Birds of the Soviet Union*. Israel Program for Science Translations, Jerusalem [translated from Russian].

Dickson, D. L., and H. G. Gilchrist. 2002. Status of marine birds of the southeastern Beaufort Sea. *Arctic* 55: 46–58.

Erskine, A. J. 1992. *Atlas of Breeding Birds of the Maritime Provinces*. Nimbus Publishing and Nova Scotia Museum, Halifax, NS (2nd ed. 2016, Bird Studies Canada, Sackville NB).

Erskine, A. J., ed. 1987. *Waterfowl Breeding Population Surveys, Atlantic Provinces*. Occasional Paper 60. Canadian Wildlife Service, Edmonton, AB.

Fay, F. H. 1961. The distribution of waterfowl to St. Lawrence Island, Alaska. *Wildfowl Trust Annual Report* 12: 70–80.

Fay, F. H., and T. J. Cade. 1959. An ecological analysis of the avifauna of St. Lawrence Island, Alaska. *University of California Publications in Zoology* 63: 73–150.

Forbes, G., K. Robertson, C. Ogilvie, and L. Seddon. 1992. Breeding densities, biogeography, and nest depredation of birds on Igloolik Island, N.W.T. *Arctic* 45: 295–303.

Forsell, D. J., and P. J. Gould. 1981. *Distribution and Abundance of Marine Birds and Mammals Wintering in the Kodiak Area of Alaska.* Biological Services Program FWS/OBS–81/13. Coastal Ecosystems Project, Office of Biological Services, US Fish and Wildlife Service, Washington, DC.

Freeman, M. M. R. 1970. The birds of the Belcher Islands, N.W.T., Canada. *Canadian Field-Naturalist* 84: 277–290.

Gaston, A. J., R. Decker, F. G. Cooch, and A. Reed. 1986. The distribution of larger species of birds breeding on the coasts of Foxe Basin and northern Hudson Bay, Canada. *Arctic* 39: 285–296.

Gauthier, J., and Y. Aubry, eds. 1996. *The Breeding Birds of Québec: Atlas of the Breeding Birds of Southern Québec.* Province of Québec Society for the Protection of Birds, Canadian Wildlife Service, Québec Region, Montréal, QC.

Gavin, A. 1947. Birds of Perry River District, Northwest Territories. *Wilson Bulletin* 59: 195–203.

Gibson, D. D. 1981. Migrant birds at Shemya Island, Aleutian Islands, Alaska. *Condor* 83: 65–77.

Gibson, D. D., and G. V. Byrd. 2007. *Birds of the Aleutian Islands, Alaska.* Nuttall Ornithological Club, Cambridge, MA, and American Ornithologists' Union, Washington, DC.

Gilchrist, H. G., and G. J. Robertson. 2000. Observations of marine birds and mammals wintering at polynyas and ice edges in the Belcher Islands, Nunavut, Canada. *Arctic* 53: 61–68.

Gill, R. E., Jr., M. R. Petersen, and P. D. Jorgensen. 1981. Birds of the north-central Alaska Peninsula, 1976–1980. *Arctic* 34: 286–306.

Gilligan, J., M. Smith, D. Rogers, and A. Contreras, eds. 1994. *Birds of Oregon: Status and Distribution.* Cinclus Publications, McMinnville, OR.

Godfrey, W. E. 1986. *The Birds of Canada.* Rev. ed. National Museum of Natural Sciences, Ottawa, ON.

Goudie, R. I., S. Brault, B. Conant, A. V. Kondratyev, M. R. Petersen, and K. Vermeer. 1994. The status of sea ducks of the North Pacific Rim: Toward their conservation and management. *Transactions of the North American Wildlife and Natural Resources Conference* 59: 27–49.

Goudie, R. I., and W. R. Whitman. 1987. Waterfowl populations in Labrador, 1980–82. Pp. 45–63 in J. A. Erskine, ed. *Waterfowl Breeding Population Surveys, Atlantic Provinces.* Occasional Paper 60. Canadian Wildlife Service, Edmonton, AB.

Groves, D. J., and E. J. Mallek. 2011. *Migratory Bird Survey in the Western Canadian Arctic, 2010.* US Fish and Wildlife Service, Juneau, AK.

Groves, D. J., and E. J. Mallek. 2012. *Migratory Bird Survey in the Western and Central Canadian Arctic, 2011.* US Fish and Wildlife Service, Juneau, AK.

Groves, D. J., E. J. Mallek, R. MacDonald, and T. J. Moser. 2009. *Migratory Bird Surveys in the Canadian Arctic, 2007.* US Fish and Wildlife Service, Juneau, AK.

Groves, D. J., E. Mallek, T. J. Moser, and L. Dickson. 2007. *Central Arctic Waterfowl Breeding Population Surveys in 2007: Sea Duck Joint Venture Progress Report.* US Fish and Wildlife Service, Washington, DC, and Canadian Wildlife Service, Ottawa.

Harris, S. W. 1966. Summer birds of the lower Kashunuk River, Yukon–Kuskokwim Delta, Alaska. *Murrelet* 47: 57–65.

Heusmann, H. W., and J. R. Sauer. 2000. The northeastern states' waterfowl breeding population survey. *Wildlife Society Bulletin* 28: 355–364.

Hodges, J. I., and W. D. Eldridge. 2001. Aerial surveys of eiders and other waterbirds on the eastern Arctic coast of Russia. *Wildfowl* 52: 127–142.

Hodges, J. I., D. Groves, and A. Breault. 2005. *Aerial survey of wintering waterbirds in the proposed Nai Kun Wind Farm Project area of Hecate Strait, 2005.* US Fish and Wildlife Service, Juneau, AK, and Canadian Wildlife Service, Delta, BC.

Hodges, J. I., D. J. Groves, and B. P. Conant. 2008. Distribution and abundance of waterbirds near shore in southeast Alaska, 1997–2002. *Northwestern Naturalist* 89: 85–96.

Hodges, J. I., J. G. King, B. Conant, and H. A. Hanson. 1996. *Aerial Surveys of Waterbirds in Alaska, 1957–94: Population Trends and Observer Variability.* Information and Technology Report 4. National Biological Service, US Department of the Interior, Juneau, AK.

Hohenberger, C. J., W. C. Hanson, and E. E. Burroughs. 1994. Birds of the Prudhoe Bay region, northern Alaska. *Western Birds* 25: 73–103.

Iverson, S. A., W. S. Boyd, H. M. Regehr, and M. S. Rodway. 2006. *Sex and Age-specific Distributions of Sea Ducks Wintering in the Strait of Georgia, British Columbia: Implications for the Use of Age Ratios as an Index of Recruitment.* Technical Report Series 459. Canadian Wildlife Service, Pacific and Yukon Region, Delta, BC.

Johnsgard, P. A. 2013. *Yellowstone Wildlife: Ecology and Natural History of the Greater Yellowstone Ecosystem.* University Press of Colorado, Boulder. 239 pp.

Johnsgard, P. A., and I. O. Buss. 1956. Waterfowl sex ratios during spring in Washington State and their interpretation. *Journal of Wildlife Management* 20:384–388.

Johnsgard, P. A., and J. Kear. 1968. A review of parental carrying of young by waterfowl. *The Living Bird* 7: 89–102.

Johnson, S. R., and D. R. Herter. 1989. *The Birds of the Beaufort Sea.* BP Exploration (Alaska), Anchorage, AK. 372 pp. (This book's literature review and bibliography of more than 1,000 references on 249 Canadian Arctic bird species is noteworthy.)

Johnson, S. R., and W. J. Richardson. 1982. Waterbird migration near the Yukon and Alaskan coast of the Beaufort Sea, 2: Moult migration of sea ducks in summer. *Arctic* 35: 291–301.

Kessel, B. 1989. *Birds of the Seward Peninsula, Alaska: Their Biogeography, Seasonality, and Natural History.* University of Alaska Press, Fairbanks, AK.

Kessel, B., and D. G. Gibson. 1978. *Status and Distribution of Alaska Birds. Studies in Avian Biology 1.* Cooper Ornithological Society, Los Angeles, CA.

Kessel, B., H. K. Springer, and C. M. White. 1964. June birds of the Kolomak River, Yukon–Kuskokwim Delta, Alaska. *Murrelet* 45: 37–47.

Kingery, H. E., ed. 1997. *Colorado Breeding Bird Atlas.* Colorado Bird Partnership, Denver.

Kistchinski, A. A. 1973. Waterfowl in northeast Asia. *Wildfowl* 24: 88–102.

Larned, W. W., R. Stehn, and R. Platte. 2011. *Waterfowl Breeding Population Survey, Arctic Coastal Plain, Alaska 2010.* Office of Migratory Bird Management, US Fish and Wildlife Service, Soldotna and Anchorage, AK.

Leopold, S. 1959. *Wildlife of Mexico: The Game Birds and Mammals.* University of California Press, Berkeley.

Lockwood, M. W., and B. Freeman. 2014. *Handbook of Texas Birds.* 2nd ed. Texas A&M University Press, College Station.

Mallek, E. J., and D. J. Groves. 2009. *Alaska–Yukon Breeding Waterfowl Population Survey: May 16 to June 7, 2009.* US Fish and Wildlife Service, Fairbanks and Juneau, AK.

Marks, J., P. Hendricks, and D. Casey. 2016. *Birds of Montana.* Buteo Books, Arlington, VA.

McLaren, P. L., and M. A. McLaren. 1982. Waterfowl populations in eastern Lancaster Sound and western Baffin Bay. *Arctic* 35: 149–157.

Merkel, F. R., A. Mosbech, D. Boertmann, and L. Grondahl. 2002. Winter seabird distribution and abundance off southwestern Greenland, 1999. *Polar Research* 21: 17–36.

Montgomerie, R. D., R. V. Carter, R. L. McLaughlin, and B. Lyon. 1983. Birds of Sarcpa Lake, Melville Peninsula, Northwest Territories: Breeding phenologies, densities and biogeography. *Arctic* 36: 65–75.

Mosbech, A., and S. Johnson. 1999. Late winter distribution and abundance of sea-associated birds in southwestern Greenland, the Davis Strait, and southern Baffin Bay. *Polar Research* 18: 1–17.

Moser, T., L. Dickson, E. Mallek, and D. Groves. 2010. *Central Arctic Waterfowl Breeding Population Surveys.* Sea Duck Joint Venture Project 98 and Arctic Goose Joint Venture Project 77. US Fish and Wildlife Service, Washington, DC, and Canadian Wildlife Service, Ottawa.

Moyle, J. B., F. B. Lee, R. L. Jessen, N. J. Ordal, R. I. Benson, J. P. Lindmeier, and L. L. Johnson. 1964. *Waterfowl in Minnesota*. Technical Bulletin 7. Division of Game and Fish, Minnesota Department of Conservation, St. Paul, MN.

Murie, O. J. 1959. *Fauna of the Aleutian Islands and Alaska Peninsula*. North American Fauna 61. US Fish and Wildlife Service, Washington, DC.

Oakleaf, B., B. Luce, S. Ritter, and A. Cerovski, eds. 1992. *Wyoming Bird and Mammal Atlas*. Wyoming Game and Fish Department, Lander.

Palmer, R. S. 1949. *Maine Birds*. Bulletin of the Museum of Comparative Zoology 102. Museum of Comparative Zoology, Cambridge, MA.

Parmelee, D. F., and S. D. MacDonald. 1960. *The Birds of West-Central Ellesmere Island and Adjacent Areas*. National Museums of Canada Bulletin 169. 10l pp.

Parmelee, D. F., H. A. Stephens, and R. H. Schmidt. 1967. *The Birds of Southeastern Victoria Island and Adjacent Small Islands*. Bulletin 222, Biological Series 78. National Museum of Canada, Ottawa.

Peck, G. K. 1972. Birds of the Cape Henrietta Maria region, Ontario. *Canadian Field-Naturalist* 86: 333–348.

Pihl, S. 1996. Western Palearctic wintering seaduck numbers. *Gibier Faune Sauvage* 13: 191–206.

Platte, R. M., and R. A. Stehn. 2009. *Abundance, Distribution, and Trend of Waterbirds on Alaska's Yukon–Kuskokwim Delta Coast, Based on 1988 to 2009 Aerial Surveys*. Division of Migratory Bird Management, US Fish and Wildlife Service, Anchorage, AK.

Portenko, L. A. 1981. *Birds of the Chukchi Peninsula and Wrangel Island*. Vol. 1. Amerind Publishing, New Delhi, India. (Translated from Russian.)

Portenko, L. A. 1989. *Birds of the Chukchi Peninsula and Wrangel Island*. Vol. 2. Smithsonian Institution Libraries and National Science Foundation, Washington, DC. (Translated from Russian.)

Preble, E. A., and W. L. McAtee. 1923. *A Biological Survey of the Pribilof Islands, Alaska: Part 1, Birds and Mammals*. North American Fauna 46. US Government Printing Office, Washington, DC.

Raftovich, R. V., S. C. Chandler, and K. A. Wilkins. 2015. *Migratory Bird Hunting Activity and Harvest during the 2013–14 and 2014–15 Hunting Seasons*. US Fish and Wildlife Service, Laurel, MD.

Raven, G. H., and D. L. Dickson. 2006. *Changes in Distribution and Abundance of Birds on Western Victoria Island from 1992–1994 to 2004–2005*. Technical Report Series 456. Canadian Wildlife Service, Ottawa.

Reed, A. 1973. *Aquatic Bird Colonies in the Saint Lawrence Estuary. Service de la Faune du Québec*. Bulletin 18. Ministère du Tourisme, de la Chasse, et de la Pêche, Québec City, QC.

Reed, A., P. Dupuis, K. Fisher, and J. Moser. 1980. *An Aerial Survey of Breeding Geese and Other Wildlife in Foxe Basin and Northern Baffin Island, Northwest Territories, July 1979*. Progress Notes 114. Canadian Wildlife Service, Ottawa.

Salomonsen, F. 1950. *The Birds of Greenland*. Part 1. Ejnar Munksgaard, Copenhagen, Denmark.

Sauer, J. R, , W. A. Link, J. E. Fallon, K. L. Pardieck, and D. J. Ziolkowski, Jr. 2013. *The North American Breeding Bird Survey 1966–2011: Summary Analysis and Species Accounts*. North American Fauna: Number 79: pp. 1–3.

Savard, J.-P. L. 1988. *A Summary of the Current Knowledge on the Distribution and Abundance of Moulting Seaducks in the Coastal Waters of British Columbia*. Technical Report Series 45. Canadian Wildlife Service, Pacific and Yukon Region, Delta, BC.

Savard, J.-P. L. 1989. Birds of rocky coastlines and pelagic waters in the Strait of Georgia. Pp. 132–141 in K. Vermeer and R. W. Butler, eds. *The Ecology and Status of Marine and Shoreline Birds in the Strait of Georgia, British Columbia*. Proceedings of a Symposium. Pacific Northwest Bird and Mammal Society and the Canadian Wildlife Service, Sidney, BC.

Savard, J.-P. L., J. Bédard, and A. Nadeau. 1999a. Spring and early summer distribution of scoters and eiders in the St. Lawrence River Estuary. Pp. 60–65 in R. I. Goudie, M. R. Petersen, and G. J. Robertson, eds. *Behaviour and Ecology of Sea Ducks*. Occasional Paper 100. Published for the Pacific Seabird Group by Canadian Wildlife Service, Ottawa.

Schiøler, E. 1925–26. *Danmarks Fugle*. 2 vol. Gyldendelske, Denmark.

Sea Duck Joint Venture. 2003 et seq. *Species Status Reports*. US Fish and Wildlife Service, Washington, DC, and Canadian Wildlife Service, Ottawa.

Sea Duck Joint Venture. 2007. *Recommendations for Monitoring Distribution, Abundance, and Trends for North American Sea Ducks*. US Fish and Wildlife Service, Washington, DC, and Canadian Wildlife Service, Ottawa.

Silverman, E., J. B. Leirness, D. T. Saalfeld, and D. Reichkus. 2012. *Atlantic Coast Wintering Sea Duck Survey, 2008–2011*. Division of Migratory Bird Management, US. Fish and Wildlife Service, Laurel, MD.

Small, A. 1994. *California Birds: Their Status and Distribution*. Ibis Publishing, Temecula, CA.

Smith, M. R., P. W. Mattocks, Jr., and K. M. Cassidy. 1997. *Breeding Birds of Washington State: Location Data and Predicted Distributions*. Vol. 4 in K. M. Cassidy, C. E. Grue, M. R. Smith, and K. M. Dvornich, eds. Washington State Gap Analysis Final Report. Seattle Audubon Society Publications in Zoology 1. Seattle Audubon Society, Seattle, WA.

Stott, R. S. 1974. *Sea Duck Populations on the New Hampshire Coastline*. Research Report 33. New Hampshire Agricultural Experimental Station, Durham.

Todd, W. E. C. 1963. *Birds of the Labrador Peninsula and Adjacent Areas*. University of Toronto Press, Toronto, ON.

Tufts, R. W. 1986. *Birds of Nova Scotia*. 3rd ed. Nimbus Publishing and Nova Scotia Museum, Halifax.

US Fish and Wildlife Service. 1999. *Population Status and Trends of Sea Ducks in Alaska*. Office of Migratory Bird Management, US Fish and Wildlife Service, Anchorage, AK.

US Fish and Wildlife Service. 2013. *North American Breeding Bird Survey, Results and Analysis 1966–2013*. Patuxent Wildlife Research Center, Laurel, MD. http://www.pwrc.usgs.gov/bbs

US Fish and Wildlife Service. 2016. *Waterfowl Population Status, 2015*. US Fish and Wildlife Service, Laurel, MD.

Veit, R. R., and W. R. Petersen. 1993. *Birds of Massachusetts*. Massachusetts Audubon Society, Lincoln, MA.

Vermeer, K., and R. W. Butler, eds. 1989. *The Ecology and Status of Marine and Shoreline Birds in the Strait of Georgia, British Columbia*. Proceedings of a Symposium. Pacific Northwest Bird and Mammal Society and the Canadian Wildlife Service, Sidney, BC.

Weller, M. W., D. L. Trauger, and G. L. Krapu. 1969. Breeding birds of the West Mirage Islands, Great Slave Lake, N.W.T. *Canadian Field-Naturalist* 83: 344–360.

Wentworth, C. 2007a. *Subsistence Migratory Bird Harvest Survey: Yukon–Kuskokwim Delta, 2001–2005*. Migratory Birds and State Programs, US Fish and Wildlife Service, Anchorage, AK.

Wentworth, C. 2007b. *Subsistence Migratory Bird Harvest Survey, Bristol Bay, 2001–2005*. Migratory Birds and State Programs, US Fish and Wildlife Service, Anchorage, AK.

Wetlands International. 2012. *Waterbird Population Estimates*. 5th ed. Wetlands International, Wageningen, The Netherlands.

Wolfe, R. J., and A. W. Paige. 1995. *The Subsistence Harvest of Black Brant, Emperor Geese and Eider Ducks in Alaska*. Division of Subsistence Paper 234. Alaska Department of Fish and Game, Juneau, AK.

Woodby, D. A., and G. J. Divoky. 1982. Spring migration of eiders and other waterbirds in Alaska. *Arctic* 35: 403–410.

Zimpfer, N. L., W. E. Rhodes, E. D. Silverman, G. S. Zimmerman, and K. D. Richkus. 2015. *Trends in Duck Breeding Populations, 1955–2015*. Administrative Report. US Fish and Wildlife Service, Laurel, MD.

Sea Duck Biology and Ecology

Afton, A. D., and S. L. Paulus. 1992. Incubation and brood care. Pp. 62–108 in B. D. J. Batt, A. D. Afton, M. G. Anderson, C. D. Ankney, D. H. Johnson, J. A. Kadlec, and G. L. Krapu, eds. *Ecology and Management of Breeding Waterfowl*. University of Minnesota Press, Minneapolis.

Agler, B. A., and S. J. Kendall. 1997. *Marine Bird and Sea Otter Population Abundance of Prince William Sound, Alaska: Trends following the T/V Exxon Valdez Oil Spill 1989–1996*. Restoration Project 96159. Office of Migratory Bird Management, US Fish and Wildlife Service, Anchorage, AK.

Alisauskas, R. T., and C. D. Ankney. 1992. The cost of egg laying and its relationship to nutrient reserves in waterfowl. Pp. 30–61 in B. D. J. Batt, A. D. Afton, M. G. Anderson, C. D. Ankney, D. H. Johnson, J. A. Kadlec, and G. L. Krapu, eds. *Ecology and Management of Breeding Waterfowl*. University of Minnesota Press, Minneapolis.

Bengtson, S.-A. 1970. Location of nest-sites of ducks in Lake Mývatn area, northeast Iceland. *Oikos* 21: 218–229.

Bengtson, S.-A. 1971a. Variation in clutch-size in ducks in relation to the food supply. *Ibis* 113: 523–526.

Bengtson, S.-A. 1971b. Habitat selection of duck broods in Lake Mývatn area, northeast Iceland. *Ornis Scandinavica* 2: 17–26.

Bengtson, S.-A. 1971c. Food and feeding of diving ducks breeding at Lake Mývatn, Iceland. *Ornis Fennica* 48: 77–92.

Boertmann, D., P. Lyngs, F. R. Merkel, and A. Mosbech. 2004. The significance of southwest Greenland as winter quarters for seabirds. *Bird Conservation International* 14: 87–112.

Brown, C. S., J. Luebbert, D. Mulcahy, J. Schamber, and D. H. Rosenberg. 2006. Blood lead levels of wild Steller's eiders (*Polysticta stelleri*) and black scoters (*Melanitta nigra*) in Alaska using a portable blood lead analyzer. *Journal of Zoo and Wildlife Medicine* 37: 361–365.

Brown, R. G. B., D. I. Gillespie, A. R. Lock, P. A. Pearce, and G. H. Watson. 1973. Bird mortality from oil slicks off eastern Canada, February–April 1970. *Canadian Field-Naturalist* 87: 225–234.

Cottam, C. 1939. *Food Habits of North American Diving Ducks*. Technical Bulletin 643. US Department of Agriculture, Washington, DC.

Décarie, R., F. Morneau, D. Lambert, S. Carrière, and J.-P. L. Savard. 1995. Habitat use by brood-rearing waterfowl in subarctic Québec. *Arctic* 48: 383–390.

Doty, H. A., F. B. Lee, and A. D. Kruse. 1975. Use of elevated nest baskets by ducks. *Wildlife Society Bulletin* 3: 68–73.

Eadie, J. M., F. P. Kehoe, and T. D. Nudds. 1988. Pre-hatch and post-hatch brood amalgamation in North American Anatidae: A review of hypotheses. *Canadian Journal of Zoology* 66: 1709–1721.

Flint, P. L., A. C. Fowler, and R. F. Rockwell. 1999. Modeling bird mortality associated with the M/V Citrus oil spill off St. Paul Island, Alaska. *Ecological Modeling* 117: 261–267.

Franson, J. C., T. E. Hollmén, P. L. Flint, J. B. Grand, and R. B. Lanctot. 2004. Contaminants in molting long-tailed ducks and nesting common eiders in the Beaufort Sea. *Marine Pollution Bulletin* 48: 504–513.

Gardarsson, A., and Á. Einarsson. 2004. Resource limitation of diving ducks at Mývatn: Food limits production. *Aquatic Ecology* 38: 285–295.

Gill, R., Jr., C. Handel, and M. Petersen. 1979. Migration of birds in Alaska marine habitats. Pp. 245–288 in *Environmental Assessment of the Alaska Continental Shelf—Final Report of Principal Investigators, Volume 5: Biological Studies*. Outer Continental Shelf Environmental Assessment Program, National Oceanic and Atmospheric Administration, Boulder, CO.

Goudie, R. I. 1984. Comparative ecology of common eiders, black scoters, oldsquaws, and harlequin ducks wintering in southeastern Newfoundland. MS thesis, University of Western Ontario, London.

Goudie, R. I. 1999. Behavior of harlequin ducks and three species of scoters wintering in the Queen Charlotte Islands, British Columbia. Pp. 6–13 in R. I. Goudie, M. R. Petersen, and G. J. Robertson, eds. *Behaviour and Ecology of Sea Ducks*. Occasional Paper 100. Canadian Wildlife Service, Ottawa.

Goudie, R. I., and C. D. Ankney. 1986. Body size, activity budgets, and diets of sea ducks wintering in Newfoundland. *Ecology* 67: 1475–1482.

Goudie, R. I., M. R. Petersen, and G. J. Robertson, eds. 1999. *Behaviour and Ecology of Sea Ducks*. Occasional Paper 100. Canadian Wildlife Service, Ottawa.

Hildén, O. 1964. Ecology of duck populations in the islands group of Valassaaret, Gulf of Bothnia. *Annales Zoologici Fennici* 1: 153–279.

Hoppe, R. T., L. M. Smith, and D. B. Wester. 1986. Foods of wintering diving ducks in South Carolina. *Journal of Field Ornithology* 57: 126–134.

Johnsgard, P. A., and I. O. Buss. 1956. Waterfowl sex ratios during spring in Washington State and their interpretation. *Journal of Wildlife Management* 20: 384–388.

King, J. G., and G. A. Sanger. 1979. Oil vulnerability index for marine oriented birds. Pp. 227–239 in J. C. Bartonek and D. N. Nettleship, eds. *Conservation of Marine Birds of Northern North America*. Research Report 11. US Fish and Wildlife Service, Washington, DC.

Kondratyev, A. V. 1999. Foraging strategies and habitat use of sea ducks breeding in northeast Russia. Pp. 52–59 in R. I. Goudie, M. R. Petersen, and G. J. Robertson, eds. *Behaviour and Ecology of Sea Ducks*. Occasional Paper 100. Canadian Wildlife Service, Ottawa.

Koskimies, J., and L. Lahti. 1964. Cold-hardiness of the newly hatched young in relation to ecology and distribution in ten species of European ducks. *Auk* 81: 281–307.

Lewis, T. L., D. Esler, and W. S. Boyd. 2007. Effects of predation by sea ducks on clam abundance in soft-bottom intertidal habitats. *Marine Ecology Progress Series* 329: 131–144.

McGilvrey, F. B. 1967. Food habits of sea ducks from the northeastern United States. *Wildfowl Trust Annual Report* 18: 142–145.

Nilsson, L. 1970. Food-seeking activity of south Swedish diving ducks in the non-breeding season. *Oikos* 21: 145–154.

Munro, J. A., and W. A. Clemens. 1931. *Water Fowl in Relation to the Spawning of Herring in British Columbia*. Bulletin 17. Biological Board of Canada, Toronto, ON.

Murdy, H. W. 1965. *Population Dynamics and Breeding Biology of Waterfowl on the Yellowknife Study Area, Northwest Territories: Annual Progress Report*. Northern Prairie Wildlife Research Center, US Fish and Wildlife Service, Jamestown, ND.

Nilsson, L. 1972. Habitat selection, food choice, and feeding habits of diving ducks in coastal waters of south Sweden during the non-breeding season. *Ornis Scandinavica* 3: 55–78.

Nyström, K. G., and O. Pehrsson. 1988. Salinity as a constraint affecting food and habitat choice of mussel-feeding diving ducks. *Ibis* 130: 94–110.

Peterson, B., and G. Gauthier. 1985. Nest site use by cavity-nesting birds of the Cariboo Parkland, British Columbia. *Wilson Bulletin* 97: 319–331.

Petrula, M. J. 1994. Nesting ecology of ducks in interior Alaska. MS thesis, University of Alaska–Fairbanks.

Piatt, J. F., C. J. Lensink, W. Butler, M. Kendziorek, and D. Nysewander. 1990. Immediate impact of the Exxon Valdez oil spill on marine birds. *Auk* 107: 387–397.

Post, E., M. C. Forchhammer, M. S. Bret-Harte, T. V. Callaghan, T. Christensen, B. Elberling, and A. D. Fox. 2009. Ecological dynamics across the Arctic associated with recent climate change. *Science* 325: 1355–1358.

Pöysä, H., and M. Milonoff. 1999. Processes underlying parental care decisions and crèching behaviour: Clarification of hypotheses. *Annales Zoologici Fennici* 36: 125–128.

Prevett, J. P., H. G. Lumsden, and F. C. Johnson. 1983. Waterfowl kill by Cree hunters of the Hudson Bay Lowland, Ontario. *Arctic* 36: 185–192.

Rohwer, F. C., and D. I. Eisenhauer. 1989. Egg mass and clutch size relationships in geese, eiders, and swans. *Ornis Scandinavica* 20: 43–48.

Rohwer, F. C., and S. Freeman. 1989. The distribution of conspecific nest parasitism in birds. *Canadian Journal of Zoology* 67: 239–253.

Schamel, D. L., D. M. Tracy, P. G. Mickelson, and A. Seguin. 1979. Avian community ecology at two sites on Espenberg Peninsula in Kotzebue Sound, Alaska. Pp. 289–607 in *Environmental Assessment of the Alaskan Continental Shelf, Volume 5: Biological Studies*. National Oceanic and Atmospheric Administration, Outer Continental Shelf Environmental Assessment Program, Boulder, CO.

Schummer, M. L., S. S. Badzinski, S. A. Petrie, Y.-W. Chen, and N. Belzile. 2010. Selenium accumulation in sea ducks wintering at Lake Ontario. *Archives of Environmental Contamination and Toxicology* 58: 854–862.

Scott-Brown, J. M. 1976. Factors affecting the habitat selection of three species of diving ducks along the Stanley Park coastline during the winter season. BS thesis, University of British Columbia, Vancouver.

Sjöberg, K. 1989. Time-related predator/prey interactions between birds and fish in a northern Swedish river. *Oecologia* 80: 1–10.

Steen, J. B., and G. W. Gabrielsen. 1986. Thermogenesis in newly hatched eider (*Somateria mollissima*) and long-tailed duck (*Clangula hyemalis*) ducklings and barnacle goose (*Branta leucopsis*) goslings. *Polar Research* 4: 181–186.

Stott, R. S., and D. P. Olson. 1972. Differential vulnerability patterns among three species of sea ducks. *Journal of Wildlife Management* 36: 775–783.

Stoudt, J. H., K. A. Trust, J. F. Cochrane, R. S. Suydam, and L. T. Quakenbush. 2002. Environmental contaminants in four eider species from Alaska and Arctic Russia. *Environmental Pollution* 119: 215–226.

Vermeer, K., and K. H. Morgan, eds. 1997. *The Ecology, Status, and Conservation of Marine and Shoreline Birds of the Queen Charlotte Islands*. Occasional Paper 93. Canadian Wildlife Service, Ottawa.

Vermeer, K., and R. C. Ydenberg. 1989. Feeding ecology of marine birds in the Strait of Georgia. Pp. 62–73 in K. Vermeer and R. W. Butler, eds. *The Ecology and Status of Marine and Shoreline Birds in the Strait of Georgia, British Columbia: Proceedings of a Symposium*. Pacific Northwest Bird and Mammal Society and Canadian Wildlife Service, Sidney, BC.

Vermeer, K., M. Bentley, K. H. Morgan, and G. E. J. Smith. 1997. Association of feeding flocks of brant and sea ducks with herring spawn at Skidegate Inlet. Pp. 102–107 in K. Vermeer and K. H. Morgan, eds. *The Ecology, Status, and Conservation of Marine and Shoreline Birds of the Queen Charlotte Islands*. Occasional Paper 93. Canadian Wildlife Service, Ottawa.

Vermeer, K., R. W. Butler, and K. H. Morgan. 1992. *The Ecology, Status, and Conservation of Marine and Shoreline Birds on the West Coast of Vancouver Island*. Occasional Paper 75. Canadian Wildlife Service, Ottawa.

Zicus, M. C., and S. K. Hennes. 1988. Cavity nesting waterfowl in Minnesota. *Wildfowl* 39: 115–123.

Steller's eider, pair

Species References

Multiple Eider Species

Abraham, K. F., and G. H. Finney. 1986. Eiders of the eastern Canadian Arctic. Pp. 55–73 in A. Reed, ed. *Eider Ducks in Canada*. Report Series 47. Canadian Wildlife Service, Ottawa.

Alexander, S. A., D. L. Dickson, and S. E. Westover. 1997. Spring migration of eiders and other waterbirds in offshore areas of the western Arctic. Pp. 6–20 in D. L. Dickson, ed. *King and Common Eiders of the Western Canadian Arctic*. Occasional Paper 94. Canadian Wildlife Service, Ottawa.

Bustnes, J. O., and K. E. Erikstad. 1988. The diets of sympatric wintering populations of common eider *Somateria mollissima* and king eider *S. spectabilis* in northern Norway. *Ornis Fennica* 65: 163–168.

Bustnes, J. O., and O. J. Lønne. 1997. Habitat partitioning among sympatric wintering common eiders *Somateria mollissima* and king eiders *Somateria spectabilis*. *Ibis* 139: 549–554.

Byers, T., and D. L. Dickson. 2001. Spring migration and subsistence hunting of king and common eiders at Holman, Northwest Territories, 1996–98. *Arctic* 54: 122–134.

Day, R. H., J. R. Rose, A. K. Prichard, R. J. Blaha, and B. A. Cooper. 2004. Environmental effects of the fall migration of eiders at Barrow, Alaska. *Marine Ornithology* 32: 13–24.

Dickson, D. L., ed. 1997. *King and Common Eiders of the Western Canadian Arctic*. Occasional Paper 94. Canadian Wildlife Service, Ottawa.

Frimer, O. 1995. Comparative behaviour of sympatric moulting populations of common eider *Somateria mollissima* and king eider *S. spectabilis* in central West Greenland. *Wildfowl* 46: 129–139.

Grand, J. B., J. C. Franson, P. L. Flint, and M. R. Petersen. 2002. Concentrations of trace elements in eggs of spectacled and common eiders on the Yukon–Kuskokwim Delta, Alaska, USA. *Environmental Toxicology and Chemistry* 21: 1673–1678.

Hamilton, D. J., and C. D. Ankney. 1994. Consumption of zebra mussels *Dreissena polymorpha* by diving ducks in Lakes Erie and St. Clair. *Wildfowl* 45: 159–166.

Hamilton, D. J., C. D. Ankney, and R. C. Bailey. 1994. Predation of zebra mussels by diving ducks: an exclosure study. *Ecology* 75: 521–531.

Johnsgard, P. A. 1964. Comparative behavior and relationships of the eiders. *Condor* 66: 113–129.

Kondratyev, A. V., and L. V. Zadorina. 1992. [Comparative ecology of the king eider *Somateria spectabilis* and spectacled eider *Somateria fischeri* on the Chaun tundra.] *Zoological Journal* 71: 99–108 (in Russian).

Myres, M. T. 1958. *Preliminary Studies of the Behavior, Migration and Distributional Ecology of Eider Ducks in Northern Alaska, 1958*. Interim progress report to the Arctic Institute of North America.

Pettingill, O. S., Jr. 1959. King eider mated with common eider in Iceland. *Wilson Bulletin* 71: 205–207.

Pettingill, O. S., Jr. 1962. A hybrid between a king eider and common eider observed in Iceland. *Wilson Bulletin* 74: 100.

Reed, A., ed. 1986. *Eider Ducks in Canada*. Report Series 47. Canadian Wildlife Service, Ottawa.

Savard, J.-P. L., J. Bédard, and A. Nadeau. 1999. Spring and early summer distribution of scoters and eiders in the St. Lawrence River Estuary. Pp. 60–65 in R. I. Goudie, M. R. Petersen, and G. J. Robertson, eds. *Behaviour and Ecology of Sea Ducks*. Occasional Paper 100. Canadian Wildlife Service, Ottawa.

Suydam, R. S., D. L. Dickson, J. B. Fadely, and L. T. Quakenbush. 2000. Population declines of king and common eiders of the Beaufort Sea. *Condor* 102: 219–222.

Suydam, R. S., L. T. Quakenbush, D. L. Dickson, and T. Obritschkewitsch. 2000. Migration of king, *Somateria spectabilis*, and common *S. mollissima vnigra* eiders past Point Barrow, Alaska, during spring and summer/fall 1996. *Canadian Field-Naturalist* 114: 444–452.

Thompson, D. Q., and R. A. Person. 1963. The eider pass at Point Barrow, Alaska. *Journal of Wildlife Management* 27: 348–356.

Wayland, M., H. G. Gilchrist, D. L. Dickson, T. Bollinger, C. James, R. A. Carreno, and J. Keating. 2001. Trace elements in king eiders and common eiders in the Canadian Arctic. *Archives of Environmental Contamination and Toxicology* 41: 491–500.

Wendt, J. S., and E. Silieff. 1986. The kill of eiders and other sea ducks by hunters in eastern Canada. Pp. 147–154 in A. Reed, ed. *Eider Ducks in Canada*. Report Series 47. Canadian Wildlife Service, Ottawa.

Woodby, D. A., and G. J. Divorky. 1982. Spring migration of eiders and other waterbirds at Point Barrow, Alaska. *Arctic* 45:403–410.

Common Eider

Ahlen, I., and A. Andersson. 1970. Breeding ecology of an eider population on Spitsbergen. *Ornis Scandinavica* 1: 83–106.

Ashcroft, R. E. 1976. A function of the pair bond in the common eider. *Wildfowl* 27: 101–105.

Baillie, S. R., and H. Milne. 1992. The influence of female age on breeding in the eider *Somateria mollissima*. *Bird Study* 29: 55–66.

Bédard, J., J.-F. Giroux, J. Huot, A. Nadeau, and B. Filion. 1999. The use of nest site shelters by common eiders in the St. Lawrence River Estuary. P. 78 in R. I. Goudie, M. R. Petersen, and G. J. Robertson, eds. *Behaviour and Ecology of Sea Ducks*. Occasional Paper 100. Published for the Pacific Seabird Group by Canadian Wildlife Service, Ottawa.

Bédard, J., J. Gauthier, and J. Munro. 1986. La distribution de l'eider à duvet durant l'élevage des canetons dan l'estuaire du Saint-Laurent. Pp. 12–19 in A. Reed, ed. *Eider Ducks in Canada*. Report Series 47. Canadian Wildlife Service, Ottawa.

Bolduc, F., and M. Guillemette. 2003. Incubation constancy and mass loss in the common eider *Somateria mollissima*. *Ibis* 145: 329–332.

Bolduc, F., M. Guillemette, and R. D. Titman. 2005. Nesting success of common eiders *Somateria mollissima* as influenced by nest-site and female characteristics in the Gulf of St. Lawrence. *Wildlife Biology* 11: 273–279.

Bordage, D., N. Plante, A. Bourget, and S. Paradis. 1998. Use of ratio estimators to estimate the size of common eider populations in winter. *Journal of Wildlife Management* 62: 185–192.

Burnett, F. L., and D. E. Snyder. 1954. Blue crab as starvation food of oiled American eiders. *Auk* 71: 315–316.

Bustnes, J. O., and K. E. Erikstad. 1990. Size selection of common mussels, *Mytilus edulis*, by common eiders, *Somateria mollissima*: energy maximization or shell weight minimization? *Canadian Journal of Zoology* 68: 2280–2283.

Bustnes, J. O., and K. E. Erikstad. 1991. Parental care in the common eider (*Somateria mollissima*): factors affecting abandonment and adoption of young. *Canadian Journal of Zoology* 69: 1538–1545.

Cantin, M., J. Bédard, and H. Milne. 1974. The food and feeding of common eiders in the St. Lawrence Estuary in summer. *Canadian Journal of Zoology* 52: 319–334.

Chapdelaine, G., A. Bourget, W. B. Kemp, D. J. Nakashima, and D. J. Murray. 1986. Population d'eider à duvet près des côtes du Québec septentrional. Pp. 39–50 in A. Reed, ed. *Eider Ducks in Canada*. Report Series 47. Canadian Wildlife Service, Ottawa.

Chaulk, K., G. J. Robertson, B. T. Collins, W. A. Montevecchi, and B. Turner. 2005. Evidence of recent population increases in common eiders breeding in Labrador. *Journal of Wildlife Management* 69: 805–809.

Chaulk, K., G. J. Robertson, and W. A. Montevecchi. 2004. Regional and annual variability in common eider nesting ecology in Labrador. *Polar Research* 23: 121–130.

Choate, J. S. 1967. Factors influencing nesting success of eiders in Penobscot Bay, Maine. *Journal of Wildlife Management* 31: 769–777.

Cooch, F. G. 1965. The breeding biology and management of the northern eider (*Somateria mollissima borealis*) in the Cape Dorset area, Northwest Territories. *Canadian Wildlife Service, Wildlife Management Bulletin*, Series 2, No. 10. 68 pp.

Cornish, B. J., and D. L. Dickson. 1997 Common eiders nesting in the western Canadian Arctic. Pp. 40–50 in D. L. Dickson, ed. *King and Common Eiders of the Western Canadian Arctic*. Occasional Paper 94. Canadian Wildlife Service, Ottawa.

Coulson, J. C. 1984. The population dynamics of the eider duck *Somateria mollissima* and evidence of extensive non-breeding by adult ducks. *Ibis* 126: 525–543.

D'Alba, L., P. Monaghan, and R. G. Nager. 2010. Advances in laying date and increasing population size suggest positive responses to climate change in common eiders *Somateria mollissima* in Iceland. *Ibis* 152: 19–28.

Dau, C. P., and W. W. Larned. 2008. *Aerial Population Survey of Common Eiders and Other Waterbirds in Near Shore Waters and along Barrier Islands of the Arctic Coastal Plain of Alaska, 24–26 June 2008.* US Fish and Wildlife Service, Anchorage, AK.

Donehower, C. E., and D. M. Bird. 2008. Gull predation and breeding success of common eiders on Stratton Island, Maine. *Waterbirds* 31: 454–462.

Donehower, C. E., and D. M. Bird. 2009. Nesting habitat use by common eiders on Stratton Island, Maine. *Wilson Journal of Ornithology* 121: 493–497.

Erikstad, K. E., J. O. Bustnes, and T. Moum. 1993. Clutch-size determination in precocial birds: A study of the common eider. *Auk* 110: 623–628.

Erikstad, K. E., T. Tveraa, and J. O. Bustnes. 1998. Significance of intraclutch egg-size variation in common eider: the role of egg size and quality of ducklings. *Journal of Avian Biology* 29: 3–9.

Erskine, A. J., and A. D. Smith. 1986. Status and movements of common eiders in the Maritime Provinces. Pp. 20–29 in A. Reed, ed. *Eider Ducks in Canada.* Report Series 47. Canadian Wildlife Service, Ottawa.

Fast, P. L. F., H. G. Gilchrist, and R. G. Clark. 2007. Experimental evaluation of nest shelter effects on weight loss in incubating common eiders *Somateria mollissima. Journal of Avian Biology* 38: 205–213.

Flint, P. L., C. L. Moran, and J. L. Schamber. 1998. Survival of common eider *Somateria mollissima* adult females and ducklings during brood rearing. *Wildfowl* 49: 103–109.

Freeman, M. M. R. 1970. Observations on the seasonal behaviour of the Hudson Bay eider (*Somateria mollissima sedentaria*). *Canadian Field-Naturalist* 84: 145–153.

Gaston, A. J., and F. G. Cooch. 1986. Observations of common eiders in Hudson Strait: aerial surveys in 1980–83. Pp. 51–54 in A. Reed, ed. *Eider Ducks in Canada.* Report Series 47. Canadian Wildlife Service, Ottawa.

Gauthier, J., and J. Bédard. 1976. Les déplacements de l'eider commun (*Somateria mollissima*) dans l'estuaire de Saint-Laurent. *Le Naturaliste Canadien* 103: 261–283.

Gauthier, J., J. Bédard, and A. Reed. 1976. Overland migration by common eiders of the St. Lawrence Estuary. *Wilson Bulletin* 88: 333–344.

Gerell, R. 1985. Habitat selection and nest predation in a common eider population in southern Sweden. *Ornis Scandinavica* 16: 129–139.

Gilliland, S. G., H. G. Gilchrist, R. F. Rockwell, G. J. Robertson, J.-P. L. Savard, F. Merkel, and A. Mosbech. 2009. Evaluating the sustainability harvest among northern common eiders *Somateria mollissima borealis* in Greenland and Canada. *Wildlife Biology* 15: 24–36.

Gorman, M. L. 1970. The daily pattern of display in a wild population of eider duck. *Wildfowl* 21: 105–107.

Gorman, M. L., and H. Milne. 1972. Crèche behaviour in the common eider *Somateria m. mollissima* L. *Ornis Scandinavica* 3: 21–26.

Gross, A. O. 1938. Eider ducks of Kent's Island. *Auk* 55: 387–400.

Gross, A. O. 1944. The present status of the American eider on the Maine coast. *Wilson Bulletin* 56: 15–26.

Guignion, D. 1969. Clutch size and incubation of the American eider (*Somateria mollissima dresseri*) on Brandypot Island. *Naturaliste Canadien*, 95: 1145–1152.

Guignion, D. L. 1967. A nesting study of the common eider (*Somateria mollissima dresseri*) in the St. Lawrence Estuary. MS thesis, Laval University, Québec City, QC.

Guillemette, M. 1994. Digestive-rate constraint in wintering common eiders (*Somateria mollissima*): implication for flying capabilities. *Auk* 111: 900–909.

Guillemette, M., J. H. Himmelman, C. Barette, and A. Reed. 1993. Habitat selection by common eiders in winter and its interaction with flock size. *Canadian Journal of Zoology* 71: 1259–1266.

Guillemette, M., D. Pelletier, J.-M. Grandbois, and P. J. Butler. 2007. Flightlessness and the energetic cost of wing molt in a large sea duck. *Ecology* 88: 2936–2945.

Guillemette, M., R. C. Ydenberg, and J. H. Himmelman. 1992. The role of energy intake rate in prey and habitat selection of common eiders *Somateria mollissima* in winter: a risk-sensitive interpretation. *Journal of Animal Ecology* 61: 599–610.

Hamilton, D. J. 1997. Community consequences of habitat use and predation by common eiders in the intertidal zone of Passamaquoddy Bay. PhD dissertation, University of Guelph, Guelph, ON.

Hamilton, D. J. 2000. Direct and indirect effects of predation by common eiders and abiotic disturbance in an intertidal community. *Ecological Monographs* 70: 21–43.

Hamilton, D. J. 2001. Feeding behavior of common eider ducklings in relation to availability of rockweed habitat and duckling age. *Waterbirds* 24: 233–241.

Hamilton, D. J., T. D. Nudds, and J. Neate. 1999. Size-selective predation of blue mussels (*Mytilus edulis*) by common eiders (*Somateria mollissima*) under controlled field conditions. *Auk* 116: 403–416.

Hanssen, S. A., H. Engebretsen, and K. E. Erikstad. 2002. Incubation start and egg size in relation to body reserve in the common eider. *Behavioral Ecology and Sociobiology* 52: 282–288.

Hario, M., and T. E. Hollmén. 2004. The role of male mate-guarding in pre-laying common eiders *Somateria m. mollissima* in the northern Baltic Sea. *Ornis Fennica* 81: 119–127.

Hario, M., M. J. Mazerolle, and P. Saurola. 2009. Survival of female common eiders *Somateria m. mollissima* in a declining population of the northern Baltic Sea. *Oecologia* 159: 747–756.

Hawkins, P. A. J., P. J. Butler, and A. J. Woakes. 1999. Growth and morphology of captive female common eider *Somateria mollissima* ducklings. *Wildfowl* 49: 92–102.

Heusmann, H. W. 1995. The eider duck. *Massachusetts Wildlife* 45(1): 31–37.

Johnson, S. R., D. R. Herter, and M. S. W. Bradstreet. 1987. Habitat use and reproductive success of Pacific eiders *Somateria mollissima vnigra* during a period of industrial activity. *Biological Conservation* 41: 77–89.

Korschgen, C. E. 1976. Breeding stress of female American eiders (*Somateria mollissima dresseri* Sharpe). PhD dissertation, University of Maine, Orono.

Korschgen, C. E. 1977. Breeding stress of female eiders in Maine. *Journal of Wildlife Management* 41: 360–373.

Korschgen, C. E., H. C. Gibbs, and H. L. Mendall. 1978. Avian cholera in eider ducks in Maine. *Journal of Wildlife Diseases* 14: 254–258.

Krementz, D. G., J. E. Hines, and D. F. Caithamer. 1996. Survival and recovery rates of American eiders in eastern North America. *Journal of Wildlife Management* 60: 855–862.

Krohn, W. B., P. O. Corr, and A. E. Hutchinson. 1992. *Status of the American Eider with Special Reference to Northern New England.* Fish and Wildlife Research 12. US Fish and Wildlife Service, Washington, DC.

Laurila, T., and M. Hario. 1988. Environmental and genetic factors influencing clutch size, egg volume, date of laying and female weight in the common eider *Somateria mollissima. Finnish Game Research* 45: 19–30.

Lusignan, A. P., K. R. Mehl, I. L. Jones, and M. L. Gloutney. 2010. Conspecific brood parasitism in common eiders (*Somateria mollissima*): Do brood parasites target safe nest sites? *Auk* 127: 765–772.

MacCharles, A. M. 1997. Diving and foraging behaviour of wintering common eiders (*Somateria mollissima*) at Cape St. Mary's, Newfoundland. MS thesis, Simon Fraser University, Burnaby, BC.

Mallory, M. L., B. M. Braune, M. Wayland, H. G. Gilchrist, and D. L. Dickson. 2004. Contaminants in common eiders (*Somateria mollissima*) of the *Canadian Arctic. Environmental Review* 12: 197–218.

Marshall, I. K. 1967. The effects of high nesting densities on the clutch size of the common eider, *Somateria mollissima* (L.). (Abstract). *Journal of Ecology* 55(3): 59.

McAloney, R. K. 1973. Brood ecology of the common eider (*Somateria mollissima dresseri*) in the Liscombe area of Nova Scotia. MS thesis, Acadia University, Wolfville, NS.

McKinney, F. 1961. An analysis of the displays of the European eider *Somateria mollissima* (Linnaeus) and Pacific eider *Somateria mollissima vnigra* Bonaparte. *Behaviour* (Supplement) 7: 1–124.

Mendall, H. L. 1968. An inventory of Maine's breeding eider ducks. *Transactions of the Northeast Section of The Wildlife Society* 25: 95–104.

Merkel, F. R. 2004. Evidence of population decline in common eiders breeding in western Greenland. *Arctic* 57: 27–36.

Milne, H. 1976. Body weights and carcass composition of the common eider. *Wildfowl* 27: 115–122.

Milne, H., and A. Reed. 1974. Annual production of fledged young from the eider colonies of the St. Lawrence Estuary. *Canadian Field-Naturalist* 88: 163–169.

Mosbech, A., G. Gilchrist, F. Merkel, C. Sonne, A. Flagstad, and H. Nyegaard. 2006. Year-round movements of northern common eiders *Somateria mollissima borealis* breeding in Arctic Canada and west Greenland followed by satellite telemetry. *Ardea* 94: 651–665.

Munro, D. A. 1961. The eider farms of Iceland. *Canadian Geographical Journal*, August 1961, 3–7.

Munro, J., and J. Bédard. 1977a. Gull predation and crèching behaviour in the common eider. *Journal of Animal Ecology* 46: 799–810.

Munro, J., and J. Bédard. 1977b. Crèche formation in the common eider. *Auk* 94: 759–771.

Nakashima, D. J. 1986. Inuit knowledge of the ecology of the common eider in northern Québec. Pp. 102–113 in A. Reed, ed. *Eider Ducks in Canada*. Report Series 47. Canadian Wildlife Service, Ottawa.

Nakashima, D. J., and D. J. Murray. 1988. *The Common Eider* (Somateria mollissima sedentaria) *of Eastern Hudson Bay: A Survey of Nest Colonies and Inuit Ecological Knowledge*. Report 102. Environmental Studies Research Funds, Ottawa, ON.

Nehls, G. 1996. Low costs of salt turnover in common eiders *Somateria mollissima*. *Ardea* 84: 23–30.

Noel, L. E., S. R. Johnson, G. M. O'Doherty, and M. K. Butcher. 2005. Common eider (*Somateria mollissima vnigrum*) nest cover and depredation on central Alaskan Beaufort Sea barrier islands. *Arctic* 58: 129–136.

Nyström, K. G., O. Pehrsson, and D. Broman. 1991. Food of juvenile common eiders (*Somateria mollissima*) in areas of high and low salinity. *Auk* 108: 250–256.

Öst, M., and M. Kilpi. 1999. Parental care influences the feeding behaviour of female eiders *Somateria mollissima*. *Annales Zoologici Fennici* 36: 195–204.

Öst, M., L. Mantila, and M. Kilpi. 2002. Shared care provides time-budgeting advantages for female eiders. *Animal Behaviour* 64: 223–231.

Parker, H., and H. Holm. 1990. Patterns of nutrient and energy expenditure in female common eiders nesting in the High Arctic. *Auk* 107: 660–668.

Paynter, R. A., Jr. 1951. Clutch-size and egg mortality of Kent Island eiders. *Ecology* 32: 497–507.

Petersen, M. R. 2009. Multiple spring migration strategies in a population of Pacific common eiders. *Condor* 111: 59–70.

Petersen, M. R., and P. L. Flint. 2002. Population structure of Pacific common eiders breeding in Alaska. *Condor* 104: 780–787.

Prach, R. W., A. R. Smith, and A. Dzubin. 1986. Nesting of the common eider near the Hell Gate-Cardigan Strait polynya, 1980–81. Pp. 127–135 in A. Reed, ed. *Eider Ducks in Canada*. Report Series 47. Canadian Wildlife Service, Ottawa.

Quinlan, S. E., and W. A. Lehnhausen. 1982. Arctic fox, *Alopex lagopus*, predation on nesting common eiders, *Somateria mollissima*, at Icy Cape, Alaska. *Canadian Field-Naturalist* 96: 462–466.

Reed, A. 1975b. Migration, homing, and mortality of breeding female eiders *Somateria mollissima dresseri* of the St. Lawrence Estuary, Québec. *Ornis Scandinavica* 6: 41–47.

Reed, A. 1986. Eiderdown harvesting and other uses of common eiders in spring and summer. Pp. 138–146 in A. Reed, ed. *Eider Ducks in Canada*. Report Series 47. Canadian Wildlife Service, Ottawa.

Reed, A., and A. J. Erskine. 1986. Populations of the common eider in eastern North America: their size and status. Pp. 156–175 in A. Reed, ed. *Eider Ducks in Canada*. Report Series 47. Canadian Wildlife Service, Ottawa.

Rigou, Y., and M. Guillemette. 2010. Foraging effort and pre-laying strategy in breeding common eiders. *Waterbirds* 33: 314–322.

Robertson, G. J., and H. G. Gilchrist. 1998. Evidence of population declines among common eiders breeding in the Belcher Islands, Northwest Territories. *Arctic* 51: 378–385.

Robertson, G. J., A. Reed, and H. G. Gilchrist. 2001. Clutch, egg, and body size variation among common eiders breeding in Hudson Bay, Canada. *Polar Research* 20: 85–94.

Robertson, G. J., M. D. Watson, and F. Cooke. 1992. Frequency, timing, and costs of intraspecific nest parasitism in the common eider. *Condor* 94: 871–879.

Savard, J.-P. L., B. Allen, D. McAuley, G. R. Milton, and S. Gilliland. 2005. Abundance and distribution of the common eider in eastern North American during the molting season [abstract]. *Second North American Sea Duck Conference, Program and Abstracts*. Patuxent Wildlife Research Center, US Geological Survey, Laurel, MD.

Schamel, D. L. 1974. The breeding biology of the Pacific eider (*Somateria mollissima vnigra* Bonaparte) on a barrier island in the Beaufort Sea, Alaska. MS thesis, University of Alaska–Fairbanks, AK.

Schamel, D. L. 1977. Breeding of the common eider (*Somateria mollissima*) on the Beaufort Sea coast of Alaska. *Condor* 79: 478–485.

Schmutz, J. K., R. J. Robertson, and F. Cooke. 1982. Female sociality in the common eider duck during brood rearing. *Canadian Journal of Zoology* 60: 3326–3331.

Schmutz, J. K., R. J. Robertson, and F. Cooke. 1983. Colonial nesting of the Hudson Bay eider duck. *Canadian Journal of Zoology* 61: 2424–2433.

Sénéchal, E., G. Gauthier, J. Bêty, H. G. Gilchrist, K. A. Hobson, and S. E. Jamieson. 2011. Do purely capital layers exist among flying birds? Evidence of exogenous contribution to arctic-nesting common eider eggs. *Oecologia* 165: 593–604.

Sonsthagen, S. A., S. L. Talbot, R. B. Lanctot, K. T. Scribner, and K. G. McCracken. 2009. Hierarchical spatial genetic structure of common eiders (*Somateria mollissima*) breeding along a migratory corridor. *Auk* 126: 744–754.

Spurr, E., and H. Milne. 1976. Adaptive significance of autumn pair formation in the common eider *Somateria mollissima* (L.). *Ornis Scandinavica* 7: 85–89.

Swennen, C. 1983. Reproductive output of eiders *Somateria m. mollissima* on the southern border of its breeding range. *Ardea* 71: 245–254.

Swennen, C. 1989. Gull predation upon eider *Somateria mollissima* ducklings: Destruction or elimination of the unfit? *Ardea* 77: 21–45.

Swennen, C. 1990. Dispersal and migratory movements of eiders *Somateria mollissima* breeding in the Netherlands. *Ornis Scandinavica* 21: 17–27.

Swennen, C., J. C. H. Ursem, and P. Duiven. 1993. Determinate laying and egg attendance in common eiders. *Ornis Scandinavica* 24: 48–52.

Swennen, C., P. Duiven, and L. A. Reyrink. 1979. Notes on the sex ratio in the common eider *Somateria mollissima* (L.). *Ardea* 67: 54–61.

van Dijk, B. 1986. The breeding biology of eiders at Île-aux-Pommes, Québec. Pp. 119–126 in A. Reed, ed. *Eider Ducks in Canada*. Report Series 47. Canadian Wildlife Service, Ottawa.

Waldeck, P., M. Kilpi, M. Öst, and M. Andersson. 2004. Brood parasitism in a population of common eider (*Somateria mollissima*). *Behaviour* 141: 725–739.

Waldeck, P., S. A. Hanssen, and M. Andersson. 2011. Indeterminate laying and flexible clutch size in a capital breeder, the common eider. *Oecologia* 165: 707–712.

Watson, M. D., G. J. Robertson, and F. Cooke. 1993. Egg-laying time and laying interval in the common eider. *Condor* 95: 869–878.

Wilson, H. M. 2007. Population ecology of Pacific common eiders on the Yukon–Kuskokwim Delta, Alaska. PhD dissertation, University of Alaska–Fairbanks.

Wilson, H. M., P. L. Flint, C. L. Moran, and A. N. Powell. 2007. Survival of breeding Pacific common eiders on the Yukon–Kuskokwim Delta, Alaska. *Journal of Wildlife Management* 71: 403–410.

King Eider

Anderson, V. R., and R. T. Alisauskas. 2001. Egg size, body size, locomotion, and feeding performance in captive king eider ducklings. *Condor* 103: 195–199.

Anderson, V. R., and R. T. Alisauskas. 2002. Composition and growth of king eider ducklings in relation to egg size. *Auk* 119: 62–70.

Bentzen, R. L., A. N. Powell, and R. S. Suydam. 2008. Factors influencing nesting success of king eiders on northern Alaska's coastal plain. *Journal of Wildlife Management* 72: 1781–1789.

Bentzen, R. L., A. N. Powell, and R. S. Suydam. 2009. Strategies for nest-site selection by king eiders. *Journal of Wildlife Management* 73: 932–938.

Conant, B., F. Roetker, and D. J. Groves. 2006. *Distribution and Abundance of Wildlife from Fixed-wing Aircraft Surveys on Victoria Island and Kent Peninsula, Nunavut, Canada*, June 2005. US Fish and Wildlife Service, Juneau, AK.

Cotter, R. C., D. L, Dickson, and C. J. Gratto. 1997. Breeding biology of the king eider in the western Canadian Arctic. Pp. 51–57 in D. L. Dickson, ed. *King and Common Eiders of the Western Canadian Arctic*. Occasional Paper 94. Canadian Wildlife Service, Ottawa.

Dau, C. P., and K. S. Bollinger. 2009. *Aerial Population Survey of Common Eiders and Other Waterbirds in Near Shore Waters and along Barrier Islands of the Arctic Coastal Plain of Alaska, 1–5 July 2009*. US Fish and Wildlife Service, Anchorage, AK.

Dickson, D. L., R. C. Cotter, J. E. Hines, and M. F. Kay. 1997. Distribution and abundance of king eiders in the western Canadian Arctic. Pp. 29–39 in D. L. Dickson, ed. *King and Common Eiders of the Western Canadian Arctic*. Occasional Paper 94. Canadian Wildlife Service, Ottawa.

Dickson, D. L., R. S. Suydam, and G. Balogh. 1999. *Tracking the Movement of King Eiders from Nesting Grounds in Western Arctic Canada to their Molting and Wintering Areas Using Satellite Telemetry: 1998/99 Progress Report*. Canadian Wildlife Service, Edmonton, AB.

Fournier, M. A., and J. E. Hines. 1994. Effects of starvation on muscle and organ mass of king eiders *Somateria spectabilis* and the ecological and management implications. *Wildfowl* 45: 188–197.

Frimer, O. 1994a. Autumn arrival and moult in king eiders (*Somateria spectabilis*) at Disko, West Greenland. *Arctic* 47: 137–141.

Frimer, O. 1994b. The behaviour of moulting king eiders *Somateria spectabilis*. *Wildfowl* 45:176–187.

Holcroft-Weerstra, A. C., and D. L. Dickson. 1997. Activity budgets of king eiders on the nesting grounds in spring. Pp. 58–66 in D. L. Dickson, ed. *King and Common Eiders of the Western Canadian Arctic*. Occasional Paper 94. Canadian Wildlife Service, Ottawa.

Kellett, D. K., and R. T. Alisauskas. 1997. Breeding biology of king eiders nesting on Karrak Lake, Northwest Territories. *Arctic* 50: 47–54.

Kellett, D. K., and R. T. Alisauskas. 2000. Body-mass dynamics of king eiders during incubation. *Auk* 117: 812–817.

Kellett, D. K., R. T. Alisauskas, and K. R. Mehl. 2003. Nest-site selection, interspecific associations, and nest success of king eiders. *Condor* 105: 373–378.

Knoche, M. J. 2004. Wing molt of king eiders: Inferences from stable isotope analyses. MS thesis, University of Alaska–Fairbanks.

Lamothe, P. 1973. Biology of the king eider (*Somateria spectabilis*) in a fresh water breeding area on Bathurst Island, N.W.T. MS thesis, University of Alberta, Edmonton.

Larned, W. W., R. Stehn, and R. Platte. 2009. *Waterfowl Breeding Population Survey, Arctic Coastal Plain, Alaska 2008*. Office of Migratory Bird Management, US Fish and Wildlife Service, Soldotna and Anchorage, AK.

Mallory, M. L., H. G. Gilchrist, S. E. Jamieson, G. J. Robertson, and D. G. Campbell. 2001. Unusual migration mortality of king eiders in central Baffin Island, Nunavut. *Waterbirds* 24: 453–456.

Mehl, K. R. 2004. Brood ecology and population dynamics of king eiders. PhD dissertation, University of Saskatchewan, Saskatoon.

Mehl, K. R., and R. T. Alisauskas. 2007. King eider (*Somateria spectabilis*) brood ecology: correlates of duckling survival. *Auk* 124: 606–618.

Mehl, K. R., R. T. Alisauskas, K. A. Hobson, and D. K. Kellett. 2004. To winter east or west? Heterogeneity in winter philopatry in a central-Arctic population of king eiders. *Condor* 106: 241–251.

Mosbech, A., and D. Boertmann. 1999. Distribution, abundance, and reaction to aerial surveys of post-breeding king eiders (*Somateria spectabilis*) in western Greenland. *Arctic* 52: 188–203.

Mosbech, A., F. Merkel, and D. Boertmann. 2002. The king eider (*Somateria spectabilis*) population in western Greenland during autumn and winter determined by aerial surveys and satellite telemetry [abstract]. *Wetlands International Seaduck Specialist Group Meeting: Workshop on Baltic/North Sea Common Eider Populations*.

Mosbech, A., R. S. Danø, F. Merkel, C. Sonne, G. Gilchrist, and A. Flagstad. 2006a. Use of satellite telemetry to locate key habitats for king eiders *Somateria spectabilis* in West Greenland. Pp. 769–776 in G. C. Boere, C. A. Galbraith, and D. A. Stroud, eds. *Waterbirds around the World*. Her Majesty's Stationery Office, Edinburgh, Scotland, UK.

Oppel, S., and A. N. Powell. 2010. Age-specific survival estimates of king eiders derived from satellite telemetry. *Condor* 112: 323–330.

Oppel, S., A. N. Powell, and D. L. Dickson. 2008. Timing and distance of king eider migration and winter movements. *Condor* 110: 296–305.

Oppel, S., A. N. Powell, and M. G. Butler. 2011. King eider foraging effort during the pre-breeding period in Alaska. *Condor* 113: 52–60.

Oppel, S., D. Dickson, and A. N. Powell. 2009. International importance of the eastern Chukchi Sea as a staging area for migrating king eiders. *Polar Biology* 32: 775–783.

Pearce, J. M., S. L. Talbot, B. J. Pierson, M. R. Petersen, K. T. Scribner, D. L. Dickson, and A. Mosbech. 2004. Lack of spatial genetic structure among nesting and wintering king eiders. *Condor* 106: 229–240.

Phillips, L. M., and A. N. Powell. 2006. Evidence for wing molt and breeding site fidelity in king eiders. *Waterbirds* 26: 148–153.

Phillips, L. M., and A. N. Powell. 2009. Brood rearing ecology of king eiders on the North Slope of Alaska. *Wilson Journal of Ornithology* 121: 430–434.

Phillips, L. M., A. N. Powell, and E. A. Rexstad. 2006. Large-scale movements and habitat characteristics of king eiders throughout the nonbreeding period. *Condor* 108: 887–900.

Phillips, L. M., A. N. Powell, E. J. Taylor, and E. A. Rexstad. 2007. Use of the Beaufort Sea by king eiders breeding on the North Slope of Alaska. *Journal of Wildlife Management* 71: 1892–1898.

Sherman, G. R. 1965. The form and duration of the male displays of the king eider (*Somateria spectabilis* Linnaeus). MS thesis, University of Nebraska, Lincoln.

Sittler, B., O. Gilg, and T. B. Berg. 2000. Low abundance of king eider nests during low lemming years in northeast Greenland. *Arctic* 53: 53–60.

Summers, R. W., L. G. Underhill, E. E. Syroechkovski, Jr., H. G. Lappo, R. P. Prys-Jones, and V. Karpov. 1994. The breeding biology of dark-bellied brent geese *Branta b. bernicla* and king eiders *Somateria spectabilis* on the northeastern Taimyr Peninsula, especially in relation to snowy owl *Nyctea scandiaca* nests. *Wildfowl* 45: 110–118.

Wayland, M., K. L. Drake, R. T. Alisauskas, D. K. Kellett, J. Traylor, C. Swoboda, and K. Mehl. 2008b. Survival rates and blood metal concentrations in two species of free-ranging North American sea ducks. *Environmental Toxicology and Chemistry* 27: 698–704.

Wayland, M., R. T. Alisauskas, D. K. Kellett, and K. R. Mehl. 2008a. Trace element concentrations in blood of nesting king eiders in the Canadian Arctic. *Archives of Environmental Contamination and Toxicology* 55: 683–690.

Spectacled Eider

Abraham, K. F. 1978. Adoption of spectacled eider ducklings by Arctic loons. *Condor* 80: 339–340.

Anderson, B. A., C. B. Johnson, B. A. Cooper, L. N. Smith, and A. A. Stickney. 1999. Habitat associations of nesting spectacled eiders on the Arctic Coastal Plain of Alaska. Pp. 27–33 in R. I. Goudie, M. R. Petersen, and G. J. Robertson, eds. *Behaviour and Ecology of Sea Ducks*. Occasional Paper 100. Published for the Pacific Seabird Group by Canadian Wildlife Service, Ottawa.

Bart, J., and S. L. Earnst. 2005. Breeding ecology of spectacled eiders *Somateria fischeri* in northern Alaska. *Wildfowl* 55: 85–100.

Dau, C. P. 1974. Nesting biology of the spectacled eider *Somateria fischeri* (Brandt) on the Yukon–Kuskokwim Delta, Alaska. MS thesis, University of Alaska–Fairbanks.

Dau, C. P., and S. A. Kistchinski. 1977. Seasonal movements and distribution of the spectacled eider. *Wildfowl* 28: 65–75.

Ely, C. R., C. P. Dau, and C. A. Babcock. 1994. Decline in a population of spectacled eiders nesting on the Yukon–Kuskokwim Delta, Alaska. *Northwestern Naturalist* 75: 81–87.

Fischer, J. B., R. A. Stehn, and G. Walters. 2009. *Nest Population Size and Potential Production of Geese and Spectacled Eiders on the Yukon–Kuskokwim Delta, Alaska, 2009*. Office of Migratory Bird Management, US Fish and Wildlife Service, Anchorage, AK.

Flint, P. L., and J. B. Grand. 1997. Survival of spectacled eider adult females and ducklings during brood rearing. *Journal of Wildlife Management* 61: 217–221.

Flint, P. L., and J. B. Grand. 1999. Incubation behavior of spectacled eiders on the Yukon–Kuskokwim Delta, Alaska. *Condor* 101: 413–416.

Flint, P. L., J. A. Morse, J. B. Grand, and C. L. Moran. 2006. Correlated growth and survival of juvenile spectacled eiders: Evidence of habitat limitation? *Condor* 108: 901–911.

Flint, P. L., J. B. Grand, J. A. Morse, and T. F. Fondell. 2000. Late summer survival of adult female and juvenile spectacled eiders on the Yukon–Kuskokwim Delta, Alaska. *Waterbirds* 23: 292–297.

Flint, P. L., M. R. Petersen, and J. B. Grand. 1997. Exposure of spectacled eiders and other diving ducks to lead in western Alaska. *Canadian Journal of Zoology* 75: 439–443.

Grand, J. B., and P. L. Flint. 1997. Productivity of nesting spectacled eiders on the lower Kashunuk River, Alaska. *Condor* 99: 926–932.

Grand, J. B., P. L. Flint, M. R. Petersen, and C. L. Moran. 1998, Effect of lead poisoning on spectacled eider survival rates. *Journal of Wildlife Management* 62: 1102–1109.

Johnsgard, P. A. 1964. Observations on the biology of the spectacled eider. *Wildfowl Trust Annual Report* 15: 104–107.

Kistchinski, A. A., and V. E. Flint. 1974. On the biology of the spectacled eider. *Wildfowl* 25: 5–15.

Lovvorn, J. R., S. E. Richman, J. M. Grebmeier, and L. W. Cooper. 2003. Diet and body condition of spectacled eiders wintering in pack ice of the Bering Sea. *Polar Biology* 26: 259–267.

Pearce, J. M., D. Esler, and A. G. Degtyarev. 1998. Nesting ecology of spectacled eiders *Somateria fischeri* on the Indigirka River Delta, Russia. *Wildfowl* 49: 110–123.

Petersen, M. R., and D. C. Douglas. 2004. Winter ecology of spectacled eiders: Environmental characteristics and population change. *Condor* 106: 79–94.

Petersen, M. R., D. C. Douglas, and D. M. Mulcahy. 1995. Use of implanted satellite transmitters to locate spectacled eiders at sea. *Condor* 97: 276–278.

Petersen, M. R., J. F. Piatt, and K. A. Trust. 1998. Foods of spectacled eiders *Somateria fischeri* in the Bering Sea, Alaska. *Wildfowl* 49: 124–128.

Petersen, M. R., W. W. Larned, and D. C. Douglas. 1999. At-sea distributions of spectacled eiders: A 120-year-old mystery resolved. *Auk* 116: 1009–1020.

Stehn, R. A., C. P. Dau, B. Conant, and W. I. Butler., Jr. 1993. Decline of spectacled eiders nesting in western Alaska. *Arctic* 46: 264–277.

Troy, D. 2003. *Molt Migration of Spectacled Eiders in the Beaufort Sea Region.* Troy Ecological Research Associates, Anchorage, AK.

US Fish and Wildlife Service. 1996a. *Spectacled Eider Recovery Plan.* US Fish and Wildlife Service, Anchorage, AK.

Steller's Eider

Bustnes, J. O. 1997. Case histories of oil spills in Norway in relation to Steller's eider. *Wetlands International Seaduck Specialist Group Bulletin* 7: 45.

Bustnes, J. O., and G. H. Systad. 2001a. Habitat use by wintering Steller's eiders *Polysticta stelleri* in northern Norway. *Ardea* 89: 267–274.

Bustnes, J. O., and G. H. Systad. 2001b. Comparative feeding ecology of Steller's eider and long-tailed ducks in winter. *Waterbirds* 24: 407–412.

Bustnes, J. O., and K. V. Galaktionov. 2004. Evidence of a state-dependent trade-off between energy intake and parasite avoidance in Steller's eiders. *Canadian Journal of Zoology* 82: 1566–1571.

Bustnes, J. O., M. Asheim, T. H. Bjørn, H. Gabrielsen, and G. H. Systad. 2000. The diet of Steller's eiders wintering in Varangerfjord, northern Norway. *Wilson Bulletin* 112: 8–13.

Dau, C. P., P. L. Flint, and M. R. Petersen. 2000. Distribution of recoveries of Steller's eiders banded on the lower Alaska Peninsula, Alaska. *Journal of Field Ornithology* 71: 541–548.

Flint, P. L., and M. P. Herzog. 1999. Breeding of Steller's eiders, *Polysticta stelleri*, on the Yukon–Kuskokwim Delta, Alaska. *Canadian Field-Naturalist* 113: 306–308.

Flint, P. L., M. R. Petersen, C. P. Dau, J. E. Hines, and J. D. Nichols. 2000. Annual survival and site fidelity of Steller's eiders molting along the Alaska Peninsula. *Journal of Wildlife Management* 64: 261–268.

Fox, A. D., and C. Mitchell. 1997. Spring habitat use and feeding behavior of Steller's eider *Polysticta stelleri* in Varangerfjord, northern Norway. *Ibis* 139: 542–548.

Fox, A. D., C. Mitchell, G. Henriksen, E. Lund, and B. Frantzen. 1997. The conservation of Steller's eider *Polysticta stelleri* at Varangerfjord, Finnmark. *Wildfowl* 48: 156–165.

Hario, M. 1997. Migration of Steller's eider in Finland. *Wetlands International Seaduck Specialist Group Bulletin* 7: 26–30.

Henriksen, G., and E. Lund. 1994. Migration times, local movements, biometric parameters, and the size and composition of the population of Steller's eider *Polysticta stelleri* in Varangerfjord in Finnmark, northern Norway. *Fauna Norvegica Series C, Cinclus* 17: 95–106.

Jones, R. D., Jr. 1965. Returns from Steller's eiders banded in Izembek Bay, Alaska. *Wildfowl Trust Annual Report* 6: 83–85.

Kertell, K. 1991. Disappearance of the Steller's eider from the Yukon–Kuskokwim Delta, Alaska. *Arctic* 44: 177–187.

Kondratyev, A. V. 1997. Overview of the status of migrating and wintering Steller's eider in east Siberia. *Wetlands International Seaduck Specialist Group Bulletin* 7: 31–34.

Larned, W. W. 1998. *Steller's eider Spring Migration Survey, Southwest Alaska, 1998.* Office of Migratory Bird Management, Waterfowl Branch, US Fish and Wildlife Service, Anchorage, AK.

Larned, W. W. 2000. *Aerial Surveys of Steller's Eiders* (Polysticta stelleri) *and Other Waterbirds and Marine Mammals in Southwest Alaska Areas Proposed for Navigation Improvements by the US Army Corps of Engineers, Alaska.* US Fish and Wildlife Service, Anchorage, AK.

Larned, W. W. 2006. *Winter Distribution and Abundance of Steller's Eiders* (Polysticta stelleri) *in Cook Inlet, Alaska, 2004–2005.* OCS Study MMS 2006–066. Office of Migratory Bird Management, Waterfowl Branch, US Fish and Wildlife Service, Anchorage, AK.

Larned, W. W. 2008. *Steller's Eider Spring Migration Surveys, Southwest Alaska, 2008.* Office of Migratory Bird Management, Waterfowl Branch, US Fish and Wildlife Service, Anchorage, AK.

Larned, W. W., and D. Zwiefelhofer. 2001. *Distribution and Abundance of Steller's Eiders* (Polysticta stelleri) *in the Kodiak Archipelago, Alaska, Jan–Feb, 2001.* Waterfowl Management, Region 7, US Fish and Wildlife Service, Soldotna, AK.

Larned, W. W., and K. S. Bollinger. 2009. *Steller's Eider Spring Migration Surveys, Southwest Alaska, 2009.* Office of Migratory Bird Management, US Fish and Wildlife Service, Anchorage, AK.

Laubhan, M. K., and K. A. Metzner. 1999. Distribution and diurnal behavior of Steller's eiders wintering on the Alaska Peninsula. *Condor* 101: 694–698.

McKinney, F. 1965. The spring behavior of wild Steller's eiders. *Condor* 67: 273–290.

Metzner, K. A. 1993. Ecological strategies of wintering Steller's eiders on Izembek Lagoon and Cold Bay, Alaska. MS thesis, University of Missouri–Columbia.

Nygård, T., B. Frantzen, and S. Švažas. 1995. Steller's eider *Polysticta stelleri* wintering in Europe: numbers, distribution and origin. *Wildfowl* 46: 140–155.

Obritschkewitsch, T., P. D. Martin, and R. S. Suydam. 2001. *Breeding Biology of Steller's Eiders Nesting near Barrow, Alaska, 1999–2000.* Technical Report NAES-TR-01-04. US Fish and Wildlife Service, Fairbanks, AK.

Obritschkewitsch, T., R. Ritchie, and J. King. 2008. *Steller's Eider Surveys near Barrow, Alaska, 2007: Final Report.* ABR Inc.—Environmental Research and Services, Fairbanks, AK.

Pearce, J. M., S. L. Talbot, M. R. Petersen, and J. R. Rearick. 2005. Limited genetic differentiation among breeding, molting, and wintering groups of the threatened Steller's eider: the role of historic and contemporary data. *Conservation Genetics* 6: 743–757.

Petersen, M. R. 1980. Observations of the wing-feather moult and summer feeding ecology of Steller's eiders at Nelson Lagoon, Alaska. *Wildfowl* 31: 99–106.

Petersen, M. R. 1981. Populations, feeding ecology and molt of Steller's eiders. *Condor* 83: 256–262.

Pihl, S. 2001. European species action plan for Steller's eider (*Polysticta stelleri*). Pp. 1–26 in N. Schäffer and U. Gallo-Orsi, eds. *European Union Action Plans for Eight Priority Bird Species.* Office for Official Publications of the European Communities, Luxembourg.

Quakenbush, L. T., and R. Suydam. 1999. Periodic nonbreeding of Steller's eiders near Barrow, Alaska, with speculation on possible causes. Pp. 34–40 in R. I. Goudie, M. R. Petersen, and G. J. Robertson, eds. *Behaviour and Ecology of Sea Ducks.* Occasional Paper 100. Published for the Pacific Seabird Group by Canadian Wildlife Service, Ottawa.

Quakenbush, L. T., R. H. Day, B. A. Anderson, F. A. Pitelka, and B. J. McCaffery. 2002. Historical and present breeding distribution of Steller's eiders in Alaska. *Western Birds* 33: 99–120.

Quakenbush, L. T., R. Suydam, T. Obritschkewitsch, and M. Deering. 2004. Breeding biology of Steller's eiders (*Polysticta stelleri*) near Barrow, Alaska, 1991–99. *Arctic* 57: 166–182.

Reed, J. A., and P. L. Flint. 2007. Movements and foraging effort of Steller's eiders and harlequin ducks wintering near Dutch Harbor, Alaska. *Journal of Field Ornithology* 78: 124–132.

Rojek, N. 2008. *Breeding Biology of Steller's Eiders Nesting near Barrow, Alaska, 2007.* Technical Report. US Fish and Wildlife Service, Fairbanks, AK.

Solovieva, D. V. 1997. Timing, habitat use and breeding biology of Steller's eider in the Lena Delta, Russia. *Wetlands International Seaduck Specialist Group Bulletin* 7: 35–39.

Solovieva, D. V., S. Phil, A. D. Fox, and J. O. Bustnes. 1998. *Polysticta stelleri* Steller's eider. *BWP Update: Journal of Birds of the Western Pacific* 2: 145–158.

Systad, G. H., and J. O. Bustnes. 2001. Coping with darkness and low temperatures: foraging strategies in Steller's eiders, *Polysticta stelleri*, wintering at high latitudes. *Canadian Journal of Zoology* 79: 402–406.

Stehn, R., and R. Platte. 2009. *Steller's Eider Distribution, Abundance, and Trend in the Arctic Coastal Plain, Alaska, 1989–2008*. Office of Migratory Bird Management, US Fish and Wildlife Service, Anchorage, AK.

Žydelis, R., and D. Esler. 2005. Response of wintering Steller's eiders to herring spawn. *Waterbirds* 28: 344–350.

Žydelis, R., S.-H. Lorentsen, A. D. Fox, A. Kuresoo, Y. Krasnov, Y. Goryaev, J. O. Bustnes, M. Hario, L. Nilsson, and A. Stipniece. 2006. Recent changes in the status of Steller's eider *Polysticta stelleri* wintering in Europe: A decline or redistribution? *Bird Conservation International* 16: 217–236.

Harlequin Duck

Adams, P. A., G. J. Robertson, and I. L. Jones. 2000. Time-activity budgets of harlequin ducks molting in the Gannet Islands, Labrador. *Condor* 102: 703–708.

Ashley, J. 1994. *Progress Report: Harlequin Duck Inventory and Monitoring in Glacier National Park, Montana*. Division of Research Management, Glacier National Park, MT.

Bengtson, S.-A. 1966. Field studies on the harlequin duck in Iceland. *Wildfowl Trust Annual Report* 17: 79–94.

Bengtson, S.-A. 1972. Breeding ecology of the harlequin duck *Histrionicus histrionicus* (L.) in Iceland. *Ornis Scandinavica* 3: 1–19.

Bengtson, S.-A., and S. Ulfstrand. 1971. Food resources and breeding frequency of the harlequin duck *Histrionicus histrionicus* in Iceland. *Oikos* 22: 235–239.

Boertmann, D. 2003. Distribution and conservation of the harlequin duck, *Histrionicus histrionicus*, in Greenland. *Canadian Field-Naturalist* 117: 249–256.

Boertmann, D., and A. Mosbech. 2002. Molting harlequin ducks in Greenland. *Waterbirds* 25: 326–332.

Bond, J. C., and D. Esler. 2006. Nutrient acquisition by female harlequin ducks prior to spring migration and reproduction: long-tailed duck and the common scoter in western Finland. *Ornis Fennica* 51: 129–145.

Bond, J. C., D. Esler, and T. D. Williams. 2008. Breeding propensity of female harlequin ducks. *Journal of Wildlife Management* 72: 1388–1393.

Bond, J. C., S. A. Iverson, N. B. MacCallum, C. M. Smith, H. J. Bruner, and D. Esler. 2009. Variation in breeding season survival of female harlequin ducks. *Journal of Wildlife Management* 73: 965–972.

Boyne, A. W. 2008. Harlequin ducks in the Canadian Maritime Provinces. *Waterbirds* 31 (Special Publication 2): 50–57.

Breault, A. [M.], and J.-P. L. Savard. 1991. *Status Report on Distribution and Ecology of Harlequin Ducks in British Columbia*. Technical Report Series 110. Canadian Wildlife Service, Pacific and Yukon Region, Delta, BC.

Brodeur, S., A. Bourget, P. Laporte, S. Marchand, G. Fitzgerald, M. Robert, and J.-P. L. Savard. 1998. *Étude des déplacements du canard arlequin* (Histrionicus histrionicus) *en Gaspésie, Québec*. Technical Report Series 331. Canadian Wildlife Service, Québec Region, Sainte-Foy, QC.

Brodeur, S., J.-P. L. Savard, M. Robert, A. Bourget, G. Fitzgerald, and R. D. Titman. 2008. Abundance and movements of harlequin ducks breeding on rivers of the Gaspé Peninsula, Québec. *Waterbirds* 31 (Special Publication 2): 122–129.

Brodeur, S., J.-P. L. Savard, M. Robert, P. Laporte, P. Lamothe, R. D. Timm, S. Marchand, S. Gilliland, and G. Fitzgerald. 2002. Harlequin duck *Histrionicus histrionicus* population structure in eastern Nearctic. *Journal of Avian Biology* 33: 127–137.

Bruner, H. J. 1997. Habitat use and productivity of harlequin ducks in the central Cascade region of Oregon. MS thesis, Oregon State University, Corvallis.

Cassirer, E. F., and C. R. Groves. 1994. *Ecology of Harlequin Ducks in Northern Idaho*. Idaho Department of Fish and Game, Boise; National Geographic Society, Washington, DC; and USDA Forest Service, Intermountain Research Station, Boise, ID.

Cassirer, E. F., C. R. Groves, and R. L. Wallen. 1991. Distribution and population status of harlequin ducks in Idaho. *Wilson Bulletin* 103: 723–725.

Cassirer, E. F., G. Schirato, F. Sharpe, C. R. Groves, and R. N. Anderson. 1993. Cavity nesting by harlequin ducks in the Pacific Northwest. *Wilson Bulletin* 105: 691–694.

Cassirer, E. F., J. D. Reichel, R. L. Wallen, and E. C. Atkinson. 1996. *Harlequin Duck* (Histrionicus histrionicus). US Forest Service/Bureau of Land Management Habitat Conservation Assessment and Conservation Strategy for the US Rocky Mountains. Idaho Department of Fish and Game, Boise.

Chubbs, T. E., B. Mactavish, and P. G. Trimper. 2000. Site characteristics of a repetitively used harlequin duck, *Histrionicus histrionicus*, nest in northern Labrador. *Canadian Field-Naturalist* 114: 324–326.

Chubbs, T. E., P. G. Trimper, G. W. Humphries, P. W. Thomas, L. T. Elson, and D. K. Laing. 2008. Tracking seasonal movements of adult male harlequin ducks from central Labrador using satellite telemetry. *Waterbirds* 31 (Special Publication 2): 173–182.

Cooke, F., G. R. Robertson, C. M. Smith, R. I. Goudie, and W. S. Boyd. 2000. Survival, emigration, and winter population structure of harlequin ducks. *Condor* 102: 137–144.

Crowley, D. W. 1994. Breeding habitat of harlequin ducks in Prince William Sound, Alaska. MS thesis, Oregon State University, Corvallis.

Crowley, D. W. 1999. Productivity of harlequin ducks breeding in Prince William Sound, Alaska. Pp. 14–20 in R. I. Goudie, M. R. Petersen, and G. J. Robertson, eds. *Behaviour and Ecology of Sea Ducks*. Occasional Paper 100. Canadian Wildlife Service, Ottawa.

Esler, D., and S. A. Iverson. 2010. Female harlequin duck winter survival 11 to 14 years after the Exxon Valdez oil spill. *Journal of Wildlife Management* 74: 471–478.

Esler, D., J. A. Schmutz, R. L. Jarvis, and D. M. Mulcahy. 2000. Winter survival of adult female harlequin ducks in relation to history of contamination by the Exxon Valdez oil spill. *Journal of Wildlife Management* 64: 839–847.

Esler, D., T. D. Bowman, K. A. Trust, B. E. Ballachey, T. A. Dean, S. C. Stephen, and C. E. O'Clair. 2002. Harlequin duck population recovery following the "Exxon Valdez" oil spill: progress, process, and constraints. *Marine Ecology Progress Series* 241: 271–286.

Fischer, J. B., and C. R. Griffin. 2000. Feeding behavior and food habits of wintering harlequin ducks at Shemya Island, Alaska. *Wilson Bulletin* 112: 318–325.

Fournier, M. A., and R. G. Bromley. 1996. Status of the harlequin duck, *Histrionicus histrionicus*, in the western Northwest Territories. *Canadian Field-Naturalist* 110: 638–641.

Gardarsson, A. 2008. Harlequin ducks in Iceland. *Waterbirds* 31 (Special Publication 2): 8–14.

Goudie, R. I. 1988. Breeding distribution of harlequin ducks in northern Labrador. *Atlantic Society of Fish and Wildlife Biologists Newsletter* 4: 17–21.

Goudie, R. I. 1989. Historical status of harlequin ducks wintering in eastern North America: a reappraisal. *Wilson Bulletin* 101: 112–114.

Goudie, R. I. 1998. *Aspects of the Ecology of Harlequin Ducks on the Torrent River, Newfoundland*. Environmental Component Study, Environmental Impact Statement for Torrent Small Hydro Company, St. Johns, NF.

Goudie, R. I. 1999. Behavior of harlequin ducks and three species of scoters wintering in the Queen Charlotte Islands, British Columbia. Pp. 6–13 in R. I. Goudie, M. R. Petersen, and G. J. Robertson, eds. *Behaviour and Ecology of Sea Ducks*. Occasional Paper 100. Canadian Wildlife Service, Ottawa.

Goudie, R. I., and I. L. Jones. 2005. Feeding behavior of harlequin ducks (*Histrionicus histrionicus*) breeding in Newfoundland and Labrador: A test of the food limitation hypothesis. *Bird Behavior* 17: 9–18.

Goudie, R. I., and S. G. Gilliland. 2008. Aspects of distribution and ecology of harlequin ducks on the Torrent River, Newfoundland. *Waterbirds* 31 (Special Publication 2): 92–103.

Goudie, R. I., D. Lemon, and J. Brazil. 1994. *Observations of Harlequin Ducks, Other Waterfowl, and Raptors in Labrador, 1987–92*. Technical Report Series 207. Canadian Wildlife Service, Atlantic Region, Sackville, NB.

Gilliland, S. G., G. J. Robertson, and G. S. Goodyear. 2008. Distribution and abundance of harlequin ducks breeding in northern Newfoundland. *Waterbirds* 31 (Special Publication 2): 104–109.

Gilliland, S. G., G. J. Robertson, M. Robert, J.-P. L. Savard, D. Amirault, P. Laporte, and P. Lamothe. 2002. Abundance and distribution of harlequin ducks molting in eastern Canada. *Waterbirds* 25: 333–339.

Gowans, B., G. J. Robertson, and F. Cooke. 1997. Behaviour and chronology of pair formation by harlequin ducks *Histrionicus histrionicus*. *Wildfowl* 48: 135–146.

Grégoire, P., J. Kneteman, and J. Allen. 1999. *Harlequin Duck Surveys in the Central Eastern Slopes of Alberta, Spring 1998*. Technical Report Series 329. Canadian Wildlife Service, Prairie and Northern Region, Edmonton, AB.

Hendricks, P. 2000. *Harlequin Duck Research and Monitoring in Montana: 1999*. Montana Natural Heritage Program, Helena, MT.

Hendricks, P., and J. D. Reichel. 1998. *Harlequin Duck Research and Monitoring in Montana: 1997*. Montana Natural Heritage Program, Helena, MT.

Hunt, W. A. 1998. The ecology of harlequin ducks (*Histrionicus histrionicus*) breeding in Jasper National Park, Canada. MS thesis, Simon Fraser University, Burnaby, BC.

Inglis, I. R., J. Lazarus, and R. Torrance. 1989. The pre-nesting behaviour and time budget of the harlequin duck *Histrionicus histrionicus*. *Wildfowl* 40: 55–73.

Iverson, S. A., and D. Esler. 2006. Site fidelity and the demographic implications of winter movements by a migratory bird, the harlequin duck *Histrionicus histrionicus*. *Journal of Avian Biology* 37: 219–228.

Iverson, S. A., and D. Esler. 2007. Survival of female harlequin ducks during wing molt. *Journal of Wildlife Management* 71: 1220–1224.

Iverson, S. A., and D. Esler. 2010. Harlequin duck population injury and recovery dynamics following the 1989 Exxon Valdez oil spill. *Ecological Applications* 20: 1993–2006.

Iverson, S. A., D. Esler, and D. J. Rizzolo. 2004. Winter philopatry of harlequin ducks in Prince William Sound, Alaska. *Condor* 106: 711–715.

Kuchel, C. R. 1977. Some aspects of the behavior and ecology of harlequin ducks breeding in Glacier National Park, Montana. MS thesis, University of Montana, Missoula.

Lazarus, J., I. Inglis, and R. L. L. F. Torrance. 2004. Mate guarding conflict, extra-pair courtship and signalling in the harlequin duck, *Histrionicus histrionicus*. *Behaviour* 141: 1061–1078.

Leafloor, J. O., J. E. Thompson, and C. D. Ankney. 1996. Body mass and carcass composition of fall migrant oldsquaws. *Wilson Bulletin* 108: 567–572.

MacCallum, B., and M. Bugera. 1998. *Harlequin Duck use of the McLeod River Watershed*. Bighorn Environmental Design, Hinton, AB.

Mallory, M. L., A. J. Fontaine, J. A. Akearok, and H. G. Gilchrist. 2008. Harlequin ducks in Nunavut. *Waterbirds* 31 (Special Issue 2): 15–18.

Mittelhauser, G. H. 2008a. Harlequin ducks in the eastern United States. *Waterbirds* 31 (Special Publication 2): 58–66.

Mittelhauser, G. H. 2008b. Apparent survival and local movements of harlequin ducks wintering at Isle au Haut. Maine. *Waterbirds* 31 (Special Publication 2): 138–146.

Mittelhauser, G. H., J. B. Drury, and P. O. Corr. 2002. Harlequin ducks (*Histrionicus histrionicus*) in Maine, 1950–1999. *Northeastern Naturalist* 9: 163–182.

Pool, W. 1962. Feeding habits of the harlequin duck. *Wildfowl Trust Annual Report* 13: 126–129.

Reed, J. A., and P. L. Flint. 2007. Movements and foraging effort of Steller's eiders and harlequin ducks wintering near Dutch Harbor, Alaska. *Journal of Field Ornithology* 78: 124–132.

Regehr, H. M. 1989. Movement patterns and population structure of harlequin ducks wintering in the Strait of Georgia, British Columbia. PhD dissertation, Simon Fraser University, Burnaby, BC.

Regehr, H. M. 2003. Survival and movement of postfledging juvenile harlequin ducks. *Wilson Bulletin* 115: 423–430.

Regehr, H. M. 2011. Movement rates and distances of wintering harlequin ducks: Implications for population structure. *Waterbirds* 34: 19–31.

Regehr, H. M., C. M. Smith, B. Arquilla, and F. Cooke. 2001. Post-fledging broods of migratory harlequin ducks accompany females to wintering areas. *Condor* 103: 408–412.

Reichel, J. D., D. L. Genter, and D. P. Hendricks. 1997. *Harlequin Duck Research and Monitoring in Montana, 1996*. Montana Natural Heritage Program, Helena.

Robert, M., and L. Cloutier. 2001. Summer food habits of harlequin ducks in eastern North America. *Wilson Bulletin* 113: 78–84.

Robert, M., G. H. Mittelhauser, B. Jobin, G. Fitzgerald, and P. Lamothe. 2008. New insights on harlequin duck population structure in eastern North America as revealed by satellite telemetry. *Waterbirds* 31 (Special Publication 2): 159–172.

Robertson, G. J., F. Cooke, R. I. Goudie, and W. S. Boyd. 1997. The timing of arrival and moult chronology of harlequin ducks *Histrionicus histrionicus*. *Wildfowl* 48: 148–155.

Robertson, G. J., F. Cooke, R. I. Goudie, and W. S. Boyd. 1998a. The timing of pair formation in harlequin ducks. *Condor* 100: 551–555.

Robertson, G. J., F. Cooke, R. I. Goudie, and W. S. Boyd. 1998b. Moult speed predicts pairing success in male harlequin ducks. *Animal Behaviour* 55: 1677–1684.

Robertson, G. J., F. Cooke, R. I. Goudie, and W. S. Boyd. 2000. Spacing patterns, mating systems, and winter philopatry in harlequin ducks. *Auk* 117: 299–307.

Robertson, G. J., G. H. Mittelhauser, T. Chubbs, P. Trimper, R. I. Goudie, P. W. Thomas, S. Brodeur, M. Robert, S. G. Gilliland, and J.-P. L. Savard. 2008. Morphological variation among harlequin ducks in the northwest Atlantic. *Waterbirds* 31 (Special Publication 2): 194–203.

Rodway, M. S. 1998. Activity patterns, diet, and feeding efficiency of harlequin ducks breeding in northern Labrador. *Canadian Journal of Zoology* 76: 902–908.

Rodway, M. S., and F. Cooke. 2002. Use of fecal analysis to determine seasonal changes in the diet of wintering harlequin ducks at a herring spawning site. *Journal of Field Ornithology* 73: 363–371.

Rodway, M. S., H. M. Regehr, and F. Cooke. 2003. Sex and age differences in distribution, abundance, and habitat preferences of wintering harlequin ducks: implications for conservation and estimating recruitment. *Canadian Journal of Zoology* 81: 492–503.

Rodway, M. S., H. M. Regehr, J. Ashley, P. V. Clarkson, R. I. Goudie, D. E. Hay, C. M. Smith, and K. G. Wright. 2003a. Aggregative response of harlequin ducks to herring spawning in the Strait of Georgia, British Columbia. *Canadian Journal of Zoology* 81: 504–514.

Rodway, M. S., J. W. Gosse, Jr., I. Fong, and W. A. Montevecchi. 1998. Discovery of a harlequin duck nest in eastern North America. *Wilson Bulletin* 110: 282–285.

Smith, C. M. 1999. *Banff National Park Harlequin Duck Research Project: 1998 Progress Report*. Heritage Resource Conservation, Parks Canada, Banff, AB.

Smith, C. M., F. Cooke, and G. J. Robertson. 2000. Long-term pair bonds in harlequin ducks. *Condor* 102: 201–205.

Smith, C. M., F. Cooke, and R. I. Goudie. 1998. Ageing harlequin duck *Histrionicus histrionicus* drakes using plumage characteristics. *Wildfowl* 49: 245–248.

Smith, C. M., R. I. Goudie, and F. Cooke. 2001. Winter age ratios and the assessment of recruitment of harlequin ducks. *Waterbirds* 24: 39–44.

Squires, K. A., K. Martin, and R. I. Goudie. 2007. Vigilance behavior in the harlequin duck (*Histrionicus histrionicus*) during the preincubation period in Labrador: Are males vigilant for self or social partner? *Auk* 124: 241–252.

Taylor, E. J. 1986. Foods and foraging ecology of oldsquaws (*Clangula hyemalis* L.) on the Arctic Coastal Plain of Alaska. MS thesis, University of Alaska–Fairbanks.

Thomas, P. W., and G. J. Robertson. 2008. Apparent survival of male harlequin ducks molting at the Gannet Islands, Labrador. *Waterbirds* 31 (Special Publication 2): 147–151.

Thomas, P. W., and M. Robert. 2001. *Updated COSEWIC Status Report, Eastern North American Harlequin Duck* (Histrionicus histrionicus). Committee on the Status of Endangered Wildlife in Canada, Ottawa, ON.

Thomas, P. W., G. H. Mittelhauser, T. E. Chubbs, P. G. Trimper, R. I. Goudie, G. J. Robertson, S. Brodeur, M. Robert, S. G. Gilliland, and J.-P. L. Savard. 2008. Movements of harlequin ducks in eastern North America. *Waterbirds* 31 (Special Publication 2): 188–193.

Torres, R., F. Cooke, G. J. Robertson, and W. S. Boyd. 2002. Pairing decisions in the harlequin duck: Costs and benefits. *Waterbirds* 25: 340–347.

Vermeer, K. 1983. Diet of the harlequin duck in the Strait of Georgia, British Columbia. *Murrelet* 64: 54–57.

Vickery, P. D. 1988. Distribution and population status of harlequin ducks (*Histrionicus histrionicus*) wintering in eastern North America. *Wilson Bulletin* 100: 119–126.

Wallen, R. L. 1987. Habitat utilization by harlequin ducks in Grand Teton National Park. MS thesis, Montana State University, Bozeman.

Long-tailed Duck (Oldsquaw)

Alison, R. 1970. The behaviour of the oldsquaw (*Clangula hyemalis*) in winter. MS thesis, University of Toronto, ON.

Alison, R. 1972. The breeding biology of the oldsquaw (*Clangula hyemalis*) at Churchill, Manitoba. PhD dissertation, University of Toronto, ON.

Alison, R. M. 1975. *Breeding Biology and Behavior of the Oldsquaw* (Clangula hyemalis L.). Ornithological Monographs 18. American Ornithologists' Union, Tampa, FL.

Alison, R. M. 1976. Oldsquaw brood behavior. *Bird-Banding* 47: 210–213.

Alison, R. M. 1977. Homing of subadult oldsquaws. *Auk* 94: 383–384.

Bergman, G. 1974. The spring migration of the long-tailed duck and the common scoter in western Finland. *Ornis Fennica* 51: 129–145.

Ellarson, R. S. 1956. A study of the old-squaw duck on Lake Michigan. PhD dissertation, University of Wisconsin, Madison.

Evans, R. M. 1970. Oldsquaws nesting in association with Arctic terns at Churchill, Manitoba. *Wilson Bulletin* 82: 383–390.

Flint, P. L., D. L. Lacroix, J. A. Reed, and R. B. Lanctot. 2004. Movements of flightless long-tailed ducks during wing molt. *Waterbirds* 27: 35–40.

Howell, M. D., J. B. Grand, and P. L. Flint. 2003. Body molt of male long-tailed ducks in the near-shore waters of the North Slope, Alaska. *Wilson Bulletin* 115: 170–175.

Jamieson, S. E., G. J. Robertson, and H. G. Gilchrist. 2001. Autumn and winter diet of long-tailed duck in the Belcher Islands, Nunavut, Canada. *Waterbirds* 24: 129–132.

Johnson, S. R. 1984. Prey selection by oldsquaws in a Beaufort Sea lagoon, Alaska. Pp. 12–19 in D. N. Nettleship, G. A. Sanger, and P. F. Springer, eds. *Marine Birds: Their Feeding Ecology and Commercial Fisheries Relationships; Proceedings of the Pacific Seabird Group Symposium, Seattle, Washington.* Canadian Wildlife Service, Dartmouth, NS.

Jones, P. H. 1979. Roosting behaviour of long-tailed ducks in relation to possible oil pollution. *Wildfowl* 30: 155–158.

Kellett, D. K., R. T. Alisauskas, K. R. Mehl, K. L. Drake, J. J. Traylor, and S. L. Lawson. 2005. Body mass of long-tailed ducks (*Clangula hyemalis*) during incubation. *Auk* 122: 313–318.

Lacroix, D. L., R. B. Lanctot, J. A. Reed, and T. L. McDonald. 2003. Effect of underwater seismic surveys on molting male long-tailed ducks in the Beaufort Sea, Alaska. *Canadian Journal of Zoology* 81: 1862–1875.

Lagler, K. F., and C. C. Wienert. 1948. Food of the old-squaw in Lake Michigan. *Wilson Bulletin* 60: 118.

Mackay, G. H. 1892. Habits of the old-squaw (*Clangula hyemalis*) in New England. *Auk* 9: 330–337.

Mallory, M. L., J. Akearok, N. R. North, D. V. Weseloh, and S. Lair. 2006. Movements of long-tailed ducks wintering on Lake Ontario to breeding areas in Nunavut, Canada. *Wilson Journal of Ornithology* 118: 494–501.

Mashima, T. Y., W. J. Fleming, and M. K. Stoskopf. 1998. Metal concentrations in oldsquaw (*Clangula hyemalis*) during an outbreak of avian cholera, Chesapeake Bay, 1994. *Ecotoxicology* 7: 107–111.

Mathiasson, S. 1970. Numbers and distribution of long-tailed ducks wintering in northern Europe. *British Birds* 63: 414–424.

Pehrsson, O. 1974. Nutrition of small ducklings regulating breeding area and reproductive output in the long-tailed duck, *Clangula hyemalis*. *International Congress of Game Biologists* 11: 259–264.

Peterson, S. R., and R. S. Ellarson. 1976. Total mercury residues in livers and eggs of oldsquaws. *Journal of Wildlife Management* 40: 704–709.

Peterson, S. R., and R. S. Ellarson. 1977. Food habits of oldsquaws wintering on Lake Michigan. *Wilson Bulletin* 89: 81–91.

Peterson, S. R., and R. S. Ellarson. 1978. P,p'-DDE, polychlorinated biphenyls, and endrin in wintering female oldsquaws in North America, 1969–73. *Pesticides Monitoring Journal* 11: 170–181.

Peterson, S. R., and R. S. Ellarson. 1979. Changes in oldsquaw carcass weight. *Wilson Bulletin* 91: 288–300.

Sanger, G. A., and R. D. Jones, Jr. 1984. Winter feeding ecology and trophic relationships of oldsquaws and white-winged scoters on Kachemak Bay, Alaska. Pages 20–28 in D. N. Nettleship, G. A. Sanger, and P. F. Springer, eds. *Marine Birds: Their Feeding Ecology and Commercial Fisheries Relationships; Proceedings of the Pacific Seabird Group Symposium, Seattle, Washington*. Canadian Wildlife Service, Dartmouth, NS.

Schamber, J. L., P. L. Flint, J. B. Grand, H. M. Wilson, and J. A. Morse. 2009. Population dynamics of long-tailed ducks breeding on the Yukon–Kuskokwim Delta, Alaska. *Arctic* 62: 190–200.

Schorger, A. W. 1947. The deep diving of the loon and old-squaw and its mechanism. *Wilson Bulletin* 59: 151–159.

Schorger, A. W. 1951. The deep diving of the old-squaw. *Auk* 63: 112.

Sea Duck Joint Venture. 2015. *Species Status Summary and Information Needs Sea Duck Joint Venture, March 2015*. Long-tailed Duck (*Clangula hyemalis*). US Fish and Wildlife Service, Washington, DC, and Canadian Wildlife Service, Ottawa.

Wilbor, S. L. 1999. *Status Report on the Bering/Pacific Oldsquaw* (Clangula hyemalis) *Population*. Alaska Natural Heritage Program, Environmental and Natural Resources Institute, University of Alaska–Anchorage.

Žydelis, R., and D. Ruskyte. 2005. Winter foraging of long-tailed ducks (*Clangula hyemalis*) exploiting different benthic communities in the Baltic Sea. *Wilson Bulletin* 117: 133–141.

Multiple Scoter Species

Anderson, E. M., J. R. Lovvorn, and M. T. Wilson. 2008. Re-evaluating marine diets of surf and white-winged scoters: Interspecific differences and the importance of soft-bodied prey. *Condor* 110: 285–295.

Bond, A. L., P. W. Hicklin, and M. Evans. 2007. Daytime spring migrations of scoters (*Melanitta* spp.) in the Bay of Fundy. *Waterbirds* 30: 566–572.

Dean, A. R., and the British Birds Rarities Committee. 1989. Distinguishing characteristics of American/East Asian race of the common scoter. *British Birds* 82: 615–616.

Dwight, J., Jr. 1914. The moults and plumages of the scoters—genus *Oidemia*. *Auk* 31: 293–398.

Gilliland, S. G., C. Lepage, J.-P. L. Savard, D. Bordage, G. J. Robertson, and E. Reed. 2009. *Developmental Surveys for Breeding Scoters in Eastern North America*. Sea Duck Joint Venture Project 115. US Fish and Wildlife Service, Washington, DC, and Canadian Wildlife Service, Ottawa.

Herter, D. R., S. M. Johnston, and A. P. Woodman. 1989. Molt migration of scoters at Cape Peirce, Alaska. *Arctic* 42: 248–252.

Lewis, T. L., D. Esler, and W. S. Boyd. 2007. Foraging behaviors of surf scoters and white-winged scoters during spawning of Pacific herring. *Condor* 109: 216–222.

Lewis, T. L., D. Esler, W. S. Boyd, and R. Žydelis. 2005. Nocturnal foraging behavior of wintering surf scoters and white-winged scoters. *Condor* 107: 637–647.

Lok, E. K., M. Kirk, D. Esler, and W. S. Boyd. 2008. Movements of pre-migratory surf and white-winged scoters in response to Pacific herring spawn. *Waterbirds* 31: 385–393.

Mallek, E. J. 2006. *Aerial Scoter and Scaup Monitoring Survey of the Yukon Flats, Alaska, 2005.* US Fish and Wildlife Service, Fairbanks, AK.

Myres, M. T. 1959. Display behavior of bufflehead, scoters, and goldeneyes at copulation. *Wilson Bulletin* 71: 159–168.

Perry, M. C., D. M. Kidwell, A. M. Wells, E. J. R. Lohnes, P. C. Osenton, and S. H. Altmann. 2006. Characterization of breeding habitats for black and surf scoters in the eastern boreal forest and subarctic regions of Canada. Pp. 80–89 in A. Hanson, J. Kerekes, and J. Paquet, eds. *Limnology and Aquatic Birds: Proceedings from the Fourth Conference, Working Group on Aquatic Birds of the Societas Internationalis Limnologiae (SIL).* Developments in Hydrobiology 189. Springer, Dordrecht, The Netherlands.

Savard, J.-P. L., and P. Lamothe. 1991. Distribution, abundance, and aspects of breeding ecology of black scoters, *Melanitta nigra,* and surf scoters, *M. perspicillata,* in northern Québec. *Canadian Field-Naturalist* 105: 488–496.

Savard, J.-P. L., J. Bédard, and A. Nadeau. 1999. Spring and early summer distribution of scoters and eiders in the St. Lawrence River Estuary. Pp. 60–65 in R. I. Goudie, M. R. Petersen, and G. J. Robertson, eds. *Behaviour and Ecology of Sea Ducks.* Occasional Paper 100. Canadian Wildlife Service, Ottawa.

Stott, R. S., and D. P. Olson. 1973. Food-habitat relationship of sea ducks on the New Hampshire coastline. *Ecology* 54: 996–1007.

Žydelis, R., D. Esler, W. S. Boyd, D. L. Lacroix, and M. Kirk. 2006. Habitat use by wintering surf and white-winged scoters: Effects of environmental attributes and shellfish aquaculture. *Journal of Wildlife Management* 70: 1754–1762.

Black and Common Scoters

Banks, A. N., W. G. Sanderson, B. Hughes, P. A. Cranswick, L. E. Smith, S. Whitehead, A. J. Musgrove, B. Haycock, and N. P. Fairney. 2008. The Sea Empress oil spill (Wales, UK): Effects on common scoter *Melanitta nigra* in Carmarthen Bay and status ten years later. *Marine Pollution Bulletin* 56: 895–902.

Bengtson, S.-A. 1966. Observations on the sexual behaviour of the common scoter (*Melanitta nigra*) on the breeding grounds, with special reference to courting parties. *Vår Fågelvärld* 25: 202–226.

Bianki, V. 1992. Seaducks of the White Sea. *IWRB* [International Waterfowl and Wetlands Research Bureau] *Seaduck Bulletin* 2: 23–29.

Décarie, R., F. Morneau, D. Lambert, S. Carrière, and J.-P. L. Savard. 1995. Habitat use by brood-rearing waterfowl in subarctic Québec. *Arctic* 48: 383–390.

Hoffman, W., and G. T. Bancroft. 1984. Molt in vagrant black scoters wintering in peninsular Florida. *Wilson Bulletin* 96: 499–504.

McKinney, F. 1959. Waterfowl at Cold Bay, Alaska, with notes on the display of the black scoter. *Wildfowl Trust Annual Report* 10: 133–140.

Ross, R. K. 1983. An estimate of the black scoter, *Melanitta nigra,* population moulting in James and Hudson Bays. *Canadian Field-Naturalist* 97: 147–150.

Sangster, G. 2009. Acoustic differences between the scoters *Melanitta nigra nigra* and *M. n. americana.* *Wilson Journal of Ornithology* 121(4): 696–702.

Sea Duck Joint Venture. 2015. *Species Status Summary and Information Needs, Sea Duck Joint Venture, March 2015. Black Scoter* (Melanitta nigra). US Fish and Wildlife Service, Washington, DC, and Canadian Wildlife Service, Ottawa.

Stehn, R., R. Platte, P. Anderson, F. Broerman, T. Moran, K. Sowl, and K. Richardson. 2006. *Monitoring Black Scoter Populations in Alaska, 2005*. Office of Migratory Bird Management, US Fish and Wildlife Service, Washington, DC.

Zalakevicius, M., and V. Jacoby. 1992. Radar and visual observations of the moult migration of common scoter within eastern Baltic. *IWRB* [International Waterfowl and Wetlands Research Bureau] *Seaduck Bulletin* 1: 51.

Surf Scoter

Gilliland, S., A. Breault, G. Dicker, K. McAloney, E. Reed, and P. Ryan. 2005. Development of methodology for capturing molting surf scoters [abstract]. Second North American Sea Duck Conference, Program and Abstracts. Patuxent Wildlife Research Center, US Geological Survey, Laurel, MD.

Henny, C. J., L. J. Blus, R. A. Grove, and S. P. Thompson. 1991. Accumulation of trace elements and organochlorines by surf scoters wintering in the Pacific Northwest. *Northwestern Naturalist* 72: 43–60.

Iverson, S. A., B. D. Smith, and F. Cooke. 2004b. Age and sex distributions of wintering surf scoters: Implications for the use of age ratios as an index of recruitment. *Condor* 106: 252–262.

Iverson, S. A., D. Esler, and W. S. Boyd. 2003. Plumage characteristics as an indicator of age class in the surf scoter. *Waterbirds* 26: 56–61.

Lacroix, D. L. 2001. Foraging impacts and patterns of wintering surf scoters feeding on bay mussels in coastal Strait of Georgia, British Columbia. MS thesis, Simon Fraser University, Burnaby, BC.

Lacroix, D. L., S. Boyd, D. Esler, M. Kirk, T. Lewis, and S. Lipovsky. 2005. Surf scoters *Melanitta perspicillata* aggregate in association with ephemerally abundant polychaetes. *Marine Ornithology* 33: 61–63.

Lesage, L., A. Reed, and J.-P. L. Savard. 1997. Plumage development and growth of wild surf scoter *Melanitta perspicillata* ducklings. *Wildfowl* 47: 198–203.

Lesage, L., A. Reed, and J.-P. L. Savard. 2008. Duckling survival and use of space by surf scoter (*Melanitta perspicillata*) broods. *Ecoscience* 15: 81–88.

Morrier, A., L. Lesage, A. Reed, and J.-P. L. Savard. 2007. *Étude sur l'Écologie de la Macreuse à Front Blanc au Lac Malbaie, Réserve des Laurentides, 1994–1995*. Technical Report Series 301. Canadian Wildlife Service, Québec Region, Québec City, QC.

Reed, A., Y. Aubry, and E. Reed. 1994. Surf scoter, *Melanitta perspicillata*, nesting in southern Québec. *Canadian Field-Naturalist* 108: 364–365.

Rosenberg, D. H., M. J. Petrula, and D. D. Hill. 2006a. *Using Satellite Telemetry to Monitor Movements of Surf Scoters* (Melanitta perspicillata) *Captured in Prince William Sound, Alaska: Final Report*. Volume 1. Exxon Valdez Oil Spill Restoration Project 273. Alaska Department of Fish and Game, Anchorage, AK.

Savard, J.-P. L., A. Reed, and L. Lesage. 1998. Brood amalgamation in surf scoters *Melanitta perspicillata* and other Mergini. *Wildfowl* 49: 129–138.

Savard, J.-P. L., A. Reed, and L. Lesage. 2007. Chronology of breeding and molt migration in surf scoters (*Melanitta perspicillata*). *Waterbirds* 30: 223–229.

Savard, J.-P. L., A. Reed, Y. Aubry, A. Morrier, L. Lesage, and E. Reed. 1999. Time budgets of surf scoter broods. Pp. 21–26 in R. I. Goudie, M. R. Petersen, and G. J. Robertson, eds. *Behaviour and Ecology of Sea Ducks*. Occasional Paper 100. Canadian Wildlife Service, Ottawa.

Vermeer, K. 1981. Food and populations of surf scoters in British Columbia. *Wildfowl* 32: 107–116.

Ward, D. H., D. J. Rizzolo, S. Herzka, M. L. Shepherd, E. Carrera-González, and D. Esler. 2005. Population delineation, winter habitat use and timing of spring migration of surf scoters from the southern portion of their winter range [abstract]. *Second North American Sea Duck Conference, Program and Abstracts*. Patuxent Wildlife Research Center, US Geological Survey, Laurel, MD.

White-winged and Velvet Scoters

Alisauskas, R. T., J. J. Traylor, C. J. Swoboda, and F. P. Kehoe. 2004. Components of population growth rate for white-winged scoters in Saskatchewan, Canada. *Animal Biodiversity and Conservation* 27 (1): 451–460.

Brown, P. W. 1977. Breeding biology of the white-winged scoter (*Melanitta fusca* deglandi). MS thesis, Iowa State University, Ames.

Brown, P. W. 1981. Reproductive ecology and productivity of white-winged scoters. PhD dissertation, University of Missouri, Columbia.

Brown, P. W., and C. S. Houston. 1982. Longevity and age of maturity of white-winged scoters. *Journal of Field Ornithology* 53: 53–54.

Brown, P. W., and L. H. Fredrickson. 1983. Growth and moult progression of white-winged scoter ducklings. *Wildfowl* 34: 115–119.

Brown, P. W., and L. H. Fredrickson. 1986. Food habits of breeding white-winged scoters. *Canadian Journal of Zoology* 64: 1652–1654.

Brown, P. W., and L. H. Fredrickson. 1987a. Body and organ weights and carcass composition of breeding female white-winged scoters. *Wildfowl* 38: 103–107.

Brown, P. W., and L. H. Fredrickson. 1987b. Time budget and incubation behavior of breeding white-winged scoters. *Wilson Bulletin* 99: 50–55.

Brown, P. W., and L. H. Fredrickson. 1989. White-winged scoter, *Melanitta fusca*, populations and nesting at Redberry Lake, Saskatchewan. *Canadian Field-Naturalist* 103: 240–247.

Brown, P. W., and M. A. Brown. 1981. Nesting biology of the white-winged scoter. *Journal of Wildlife Management* 45: 38–45.

Dobush, G. R. 1986. The accumulation of nutrient reserves and their contribution to reproductive success in the white-winged scoter. MS thesis, University of Guelph, Guelph, ON.

Duebbert, H. F. 1961. Recent brood records for white-winged scoter in North Dakota. *Wilson Bulletin* 73: 209–210.

Grosz, T., and C. F. Yocom. 1972. Food habits of the white-winged scoter in northwestern California. *Journal of Wildlife Management* 36: 1279–1282.

Kehoe, F. P. 1989. The adaptive significance of crèching behavior in the white-winged scoter (*Melanitta fusca deglandi*). *Canadian Journal of Zoology* 67: 406–411.

Kehoe, F. P., P. W. Brown, and C. S. Houston. 1989. Survival and longevity of white-winged scoters nesting in central Saskatchewan. *Journal of Field Ornithology* 60: 133–136.

Krementz, D. G., P. W. Brown, F. P. Kehoe, and C. S. Houston. 1997. Population dynamics of white-winged scoters. *Journal of Wildlife Management* 61: 222–227.

Koskimies, J. 1955. Juvenile mortality and population balance in the velvet scoter (*Melanitta fusca*) in maritime conditions. *Acta XI International Ornithological Congress* 1954, pp. 476–479.

Koskimies, J. 1957. Nistortreue and Sterblichkeit bei einem marinen Bestand der Samtente. *Vogelwarte* 18: 46–51. (Velvet scoter nest fidelity and mortality.)

Koskimies, J. 1958. [Juvenile mortality and its causes in the velvet scoter (*Melanitta fusca*)]. *Suomen Riista* 12: 70–88. (In Finnish, with English summary.)

Koskimies, J., and E. Routamo. 1953. Zur Fortpfianzungsbiologie der Samtente *Melanitta f. fusca* (L). 1. Allgemeine Nistokologie. *Papers on Game Research (Riistatietelliaia Julkaisuja)* 10: 1–105. (Breeding biology of the velvet scoter.)

Rawls, C. K., Jr. 1949. An investigation of the life history of the white-winged scoter (*Melanitta fusca deglandi*). MS thesis, University of Minnesota–Minneapolis.

Richman, S. E., and J. R. Lovvorn. 2008. Costs of diving by wing and foot propulsion in a sea duck, the white-winged scoter. *Journal of Comparative Physiology B* 178: 321–332.

Rosenberg, D. H., M. J. Petrula, and D. D. Hill. 2006. *Seasonal Movements of White-winged Scoters* (Melanitta fusca) *from Prince William Sound, Alaska: Final Report.* Volume 2. Exxon Valdez Oil Spill Restoration Project 273. Alaska Department of Fish and Game, Anchorage.

Safine, D. E. 2005. Breeding ecology of white-winged scoters on the Yukon Flats, Alaska. MS thesis, University of Alaska–Fairbanks.

Safine, D. E., and M. S. Lindberg. 2008. Nest habitat selection of white-winged scoters on Yukon Flats, Alaska. *Wilson Journal of Ornithology* 120: 582–593.

Sanger, G. A., and R. D. Jones, Jr. 1984. Winter feeding ecology and trophic relationships of oldsquaws and white-winged scoters on Kachemak Bay, Alaska. Pp. 20–28 in D. N. Nettleship, G. A. Sanger, and P. F. Springer, eds. *Marine Birds: Their Feeding Ecology and Commercial Fisheries Relationships; Proceedings of the Pacific Seabird Group Symposium, Seattle, Washington.* Canadian Wildlife Service, Dartmouth, NS.

Sea Duck Joint Venture. 2015. *Species Status Summary and Information Needs, Sea Duck Joint Venture, March 2015. White-winged Scoter* (Melanitta fusca). US Fish and Wildlife Service, Washington, DC, and Canadian Wildlife Service, Ottawa.

Traylor, J. J., and R. T. Alisauskas. 2006. Effects of intrinsic and extrinsic factors on survival of white-winged scoter (*Melanitta fusca deglandi*) ducklings. *Auk* 123: 67–81.

Traylor, J. J., R. T. Alisauskas, and F. P. Kehoe. 2004a. Nesting ecology of white-winged scoters (*Melanitta fusca deglandi*) at Redberry Lake, Saskatchewan. *Auk* 121: 950–962.

Traylor, J. J., R. T. Alisauskas, and F. P. Kehoe. 2004b. Multistate modeling of brood amalgamation in white-winged scoters *Melanitta fusca deglandi. Animal Biodiversity and Conservation* 27: 369–370.

Traylor, J. J., R. T. Alisauskas, and F. P. Kehoe. 2008. Ecological correlates of duckling adoption among white-winged scoters *Melanitta fusca*: Strategy, epiphenomenon, or combination? *Behavioral Ecology and Sociobiology* 62: 1085–1097.

Vermeer, K. 1969. Some aspects of the breeding of the white-winged scoter at Miquelon Lake, Alberta. *Blue Jay* 27: 72–73.

Vermeer, K., and N. Bourne. 1984. The white-winged scoter diet in British Columbia waters: Resource partitioning with other scoters. Pp. 30–38 in D. N. Nettleship, G. A. Sanger, and P. F. Springer, eds. *Marine Birds: Their Feeding Ecology and Commercial Fisheries Relationships: Proceedings of the Pacific Seabird Group Symposium, Seattle, Washington.* Pacific Seabird Group Symposium. Minister of Supply and Services Canada, Ottawa, ON.

Multiple *Bucephala* Species

Bourget, D., J.-P. L. Savard, and M. Guillemette. 2007. Distribution, diet and dive behavior of Barrow's and common goldeneyes during spring and autumn in the St. Lawrence Estuary. *Waterbirds* 30: 230–240.

Campbell, J. M. 1969. The canvasback, common goldeneye, and bufflehead in arctic Alaska. *Condor* 71: 80.

Carney, S. S. 1983. Species, age, and sex identification of Nearctic goldeneyes from wings. *Journal of Wildlife Management* 47: 754–761.

Eadie, J. M. 1989. Alternative reproductive tactics in a precocial bird: The ecology and evolution of brood parasitism in goldeneyes. PhD dissertation, University of British Columbia, Vancouver.

Eadie, J. M., and B. E. Lyon. 1998. Cooperation, conflict, and crèching behavior in goldeneye ducks. *American Naturalist* 151: 397–408.

Eadie, J. M., and G. Gauthier. 1985. Prospecting for nest sites by cavity-nesting ducks of the genus *Bucephala. Condor* 87: 528–534.

Eadie, J. M., and H. G. Lumsden. 1985. Is nest parasitism always deleterious to goldeneyes? *American Naturalist* 126: 859–866.

Eadie, J. M., and J. Fryxell. 1992. Density-dependence, frequency dependence, and alternative nesting strategies in goldeneyes. *American Naturalist* 140: 621–641.

Eadie, J. M., J. N. M. Smith, D. Zadworny, U. Kühnlein, and K. Cheng. 2010. Probing parentage in parasitic birds: an evaluation of methods to detect conspecific brood parasitism using goldeneyes *Bucephala islandica* and *B. clangula* as a test case. *Journal of Avian Biology* 41: 163–176.

Eadie, J. M., P. W. Sherman, and B. Semel. 1998. Conspecific brood parasitism, population dynamics, and the conservation of cavity-nesting birds. Pp. 306–340 in T. M. Caro, ed. *Behavioral Ecology and Conservation Biology*. Oxford University Press, Oxford, England. UK.

Erskine, A. J. 1990. Joint laying in *Bucephala* ducks: "Parasitism" or nest-site competition? *Ornis Scandinavica* 21: 52–56.

Evans, M. R., D. B. Lank, W. S. Boyd, and F. Cooke. 2002. A comparison of the characteristics and fate of Barrow's goldeneye and bufflehead nests in nest boxes and natural cavities. *Condor* 104: 610–619.

Fitzner, R. E., and R. H. Gray. 1994. Winter diet and weights of Barrow's and common goldeneyes in south-central Washington. *Northwest Science* 68: 172–177.

Gauthier, G. 1987. Further evidence of long-term pair bonds in ducks of the genus *Bucephala*. *Auk* 104: 521–522.

Lavers, J. L., J. E. Thompson, C. A. Paszowski, and C. D. Ankney. 2006. Variation in size and composition of bufflehead (*Bucephala albeola*) and Barrow's goldeneye (*Bucephala islandica*) eggs. *Wilson Journal of Ornithology* 118: 173–177.

Lumsden, H. G., G. J. Robinson, and R. Hartford. 1986. Choice of nest boxes by cavity-nesting ducks. *Wilson Bulletin* 98: 167–168.

Mallory, M. L., and P. J. Weatherhead. 1990. Effects of nest parasitism and nest location on eggshell strength in waterfowl. *Condor* 92: 1031–1039.

Martin, P. R., and B. M. Di Labio. 1994. Natural hybrids between the common goldeneye, *Bucephala clangula*, and Barrow's goldeneye, *B. islandica*. *Canadian Field-Naturalist* 108: 195–198.

Myres, M. T. 1959b. Display behavior of bufflehead, scoters, and goldeneyes at copulation. *Wilson Bulletin* 71: 159–168.

Ouellet, J.-F., M. Guillemette, and M. Robert. 2010. Spatial distribution and habitat selection of Barrow's and common goldeneyes wintering in the St. Lawrence marine system. *Canadian Journal of Zoology* 88: 306–314.

Savard, J.-P. L. 1982. Intra- and inter-specific competition between Barrow's goldeneye (*Bucephala islandica*) and bufflehead (*Bucephala albeola*). *Canadian Journal of Zoology* 60: 3439–3446.

Savard, J.-P. L. 1984. Territorial behaviour of common goldeneye, Barrow's goldeneye and bufflehead in areas of sympatry. *Ornis Scandinavica* 15: 211–216.

Savard, J.-P. L. 1987. Causes and functions of brood amalgamation in Barrow's goldeneye and bufflehead. *Canadian Journal of Zoology* 65: 1548–1553.

Savard, J.-P. L., and J. M. Eadie. 1989. Survival and breeding philopatry in Barrow's and common goldeneyes. *Condor* 91: 198–203.

Savard, J.-P. L., G. E. J. Smith, and J. N. M. Smith. 1991. Duckling mortality in Barrow's goldeneye and bufflehead broods. *Auk* 108: 568–577.

Thompson, J. E. 1996. Comparative reproductive ecology of female buffleheads (*Bucephala albeola*) and Barrow's goldeneye (*Bucephala islandica*) in central British Columbia. PhD dissertation, University of Western Ontario, London, ON.

Thompson, J. E., and C. D. Ankney. 2002. Role of food in territoriality and egg production of buffleheads (*Bucephala albeola*) and Barrow's goldeneyes (*Bucephala islandica*). *Auk* 119: 1075–1090.

Tobish, T. 1986. Separation of Barrow's and common goldeneyes in all plumages. *Birding* 18: 17–30.

Vermeer, K. 1982. Food and distribution of three *Bucephala* species in British Columbia waters. *Wildfowl* 33: 22–30.

Bufflehead

Donaghey, R. H. 1975. Spacing behaviour of breeding buffleheads (*Bucephala albeola*) on ponds in the southern boreal forest. MS thesis, University of Alberta, Edmonton, AB.

Erskine, A. J. 1960. A discussion of the distributional ecology of the bufflehead (*Bucephala albeola*; Anatidae: Aves) based on breeding biology studies in British Columbia. MS thesis, University of British Columbia, Vancouver.

Erskine, A. J. 1961. Nest-site tenacity and homing in the bufflehead. *Auk* 78: 389–396.

Erskine, A. J. 1972c. *Buffleheads*. Monograph Series 4. Canadian Wildlife Service, Ottawa.

Gauthier, G. 1986. Experimentally induced polygyny in buffleheads: Evidence for a mixed reproductive strategy? *Animal Behaviour* 34: 300–302.

Gauthier, G. 1987b. The adaptive significance of territorial behaviour in breeding buffleheads: A test of three hypotheses. *Animal Behaviour* 35: 348–360.

Gauthier, G. 1987c. Brood territories in buffleheads: Determinants and correlates of territory size. *Canadian Journal of Zoology* 65: 1402–1410.

Gauthier, G. 1988. Factors affecting nest-box use by buffleheads and other cavity-nesting birds. *Wildlife Society Bulletin* 16: 132–141.

Gauthier, G. 1989. The effect of experience and timing on reproductive performance in buffleheads. *Auk* 106: 568–576.

Gauthier, G. 1990. Philopatry, nest-site fidelity, and reproductive performance in buffleheads. *Auk* 107: 126–132.

Gauthier, G., and J. N. M. Smith. 1987. Territorial behavior, nest-site availability, and breeding density in buffleheads. *Journal of Animal Ecology* 56: 171–184.

Henny, C. J., J. L. Carter, and B. J. Carter. 1981. A review of bufflehead sex and age criteria with notes on weights. *Wildfowl* 32: 117–122.

Knutsen, G. A., and J. C. King. 2004. Bufflehead breeding activity in south-central North Dakota. *Prairie Naturalist* 36: 187–190.

Limpert, R. J. 1980. Homing success of adult buffleheads to a Maryland wintering site. *Journal of Wildlife Management* 44: 905–908.

Wienmeyer, S. N. 1967. Bufflehead food habits, parasites, and biology in northern California. MS thesis, Humboldt State College, Arcata, CA.

Barrow's Goldeneye

Daury, R. W., and M. C. Bateman. 1996. *The Barrow's Goldeneye* (Bucephala islandica) *in the Atlantic Provinces and Maine*. Canadian Wildlife Service, Atlantic Region, Sackville, NB.

Edwards, R. Y. 1953. Barrow's goldeneye using crow nests in British Columbia. *Wilson Bulletin* 65: 197–198.

Einarsson, Á. 1990. Settlement into breeding habitats by Barrow's goldeneyes *Bucephala islandica*: Evidence for temporary oversaturation of preferred habitat. *Ornis Scandinavica* 21: 7–16.

Harris, S. W., C. L. Buechele, and C. F. Yocom. 1954. The status of Barrow's goldeneye in eastern Washington. *Murrelet* 35: 33–38.

Hobson, K. A., J. E. Thompson, M. R. Evans, and S. Boyd. 2005. Tracing nutrient allocation to reproduction in Barrow's goldeneye. *Journal of Wildlife Management* 69: 1221–1228.

Hogan, D., J. E. Thompson, D. Esler, and W. S. Boyd. Discovery of important postbreeding sites for Barrow's goldeneye in the Boreal Transition Zone of Alberta. *Waterbirds* 34: 261–388.

Jaatinen, K., S. Jaari, R. B. O'Hara, M. Öst, and J. Merilä. 2009. Relatedness and spatial proximity as determinants of host-parasite interactions in the brood parasitic Barrow's goldeneye (*Bucephala islandica*). *Molecular Ecology* 18: 2713–2721.

Koehl, P. S., T. C. Rothe, and D. V. Derksen. 1982. Winter food habits of Barrow's goldeneyes in southeast Alaska. Pp. 1–5 in D. N. Nettleship, G. A. Sanger, and P. F. Springer, eds. *Marine Birds: Their Feeding Ecology and Commercial Fisheries Relationships; Proceedings of the Pacific Seabird Group Symposium, Seattle, Washington*. Canadian Wildlife Service, Dartmouth, NS.

Munro, J. A. 1939. Studies of waterfowl in British Columbia: Barrow's golden-eye, American golden-eye. *Transactions of the Royal Canadian Institute* 22: 259–318.

Robert, M., and J.-P. L. Savard. 2006. The St. Lawrence River Estuary and Gulf: A stronghold for Barrow's goldeneyes wintering in eastern North America. *Waterbirds* 29: 437–450.

Robert, M., B. Drolet, and J.-P. L. Savard. 2008. Habitat features associated with Barrow's goldeneye breeding in eastern Canada. *Wilson Journal of Ornithology* 120: 320–330.

Robert, M., D. Bordage, J.-P. L. Savard, G. Fitzgerald, and F. Morneau. 2000. The breeding range of the Barrow's goldeneye in eastern North America. *Wilson Bulletin* 112: 1–7.

Robert, M., M.-A. Vaillancourt, and P. Drapeau. 2010. Characteristics of nest cavities of Barrow's goldeneyes in eastern Canada. *Journal of Field Ornithology* 81: 287–293.

Robert, M., R. Benoit, and J.-P. L. Savard. 2002. Relationship among breeding, molting, and wintering areas of male Barrow's goldeneyes (*Bucephala islandica*) in eastern North America. *Auk* 119: 676–684.

Savard, J.-P. L. 1985. Evidence of long-term pair bonds in Barrow's goldeneye (*Bucephala islandica*). *Auk* 102: 389–390.

Savard, J.-P. L. 1987. *Status Report on Barrow's Goldeneye*. Technical Report Series 23. Canadian Wildlife Service, Pacific and Yukon Region, Delta, BC.

Savard, J.-P. L. 1988a. Winter, spring and summer territoriality in Barrow's goldeneyes: characteristics and benefits. *Ornis Scandinavica* 19: 119–128.

Savard, J.-P. L. 1988b. Use of nest boxes by Barrow's goldeneyes: Nesting success and effect on the breeding population. *Wildlife Society Bulletin* 16: 125–131.

Savard, J.-P. L., and P. Dupuis. 1999. A case for concern: The eastern population of Barrow's goldeneye. Pp. 66–76 in R. I. Goudie, M. R. Petersen, and G. J. Robertson, eds. *Behaviour and Ecology of Sea Ducks*. Occasional Paper 100. Canadian Wildlife Service, Ottawa.

Savard, J.-P. L., and J. N. M. Smith. 1987. Interspecific aggression by Barrow's goldeneye: a descriptive and functional analysis. *Behaviour* 102: 168–184.

Sawyer, E. J. 1928. The courtship behavior of Barrow's goldeneye (*Glaucionetta islandica*). *Wilson Bulletin* 40: 5–17.

Schmelzer, I. 2006. *A Management Plan for Barrow's Goldeneye (*Bucephala islandica; *Eastern Population) in Newfoundland and Labrador*. Wildlife Division, Department of Environment and Conservation, Corner Brook, NF.

van de Wetering, D. 1997. *Moult Characteristics and Habitat Selection of Postbreeding Male Barrow's Goldeneye (*Bucephala islandica) *in Northern Yukon*. Technical Report Series 296. Canadian Wildlife Service, Pacific and Yukon Region, Nepean, ON.

van de Wetering, D., and F. Cooke. 2000. Body weight and feather growth of male Barrow's goldeneye during wing molt. *Condor* 102: 228–231.

Common Goldeneye

Afton, A. D., and R. D. Sayler. 1982. Social courtship and pair-bonding of common goldeneyes, *Bucephala clangula*, wintering in Minnesota. *Canadian Field-Naturalist* 96: 295–300.

Andersson, M., and M. O. G. Eriksson. 1982. Nest parasitism in goldeneyes *Bucephala clangula:* some evolutionary aspects. *American Naturalist* 120: 1–16.

Breckenridge, W. J. 1953. Night rafting of American golden-eyes on the Mississippi River. *Auk* 70: 201–204.

Carter, B. C. 1958. *The American Goldeneye in Central New Brunswick*. Management Bulletin Series 2, Number 9. Canadian Wildlife Service Wildlife, Ottawa.

Coulter, M. W., W. Crenshaw, G. Donovan, and J. Dorso. 1979. An experiment to establish a goldeneye population. *Wildlife Society Bulletin* 7: 116–118.

Dane, B., and W. G. van der Kloot. 1964. An analysis of the display of the goldeneye duck, *Bucephala clangula* (L.) *Behaviour* 22: 282–328.

Dane, B., C. Walcott, and W. H. Drury. 1959. The form and duration of the display actions of the goldeneye (*Bucephala clangula*). *Behaviour* 14: 265–281.

Dennis, R. H., and H. Dow. 1984. The establishment of a population of goldeneyes *Bucephala clangula* breeding in Scotland. *Bird Study* 31: 217–222.

Dow, H., and S. Fredga. 1983. Breeding and natal dispersal of the goldeneye, *Bucephala clangula*. *Journal of Animal Ecology* 52: 681–695.

Dow, H., and S. Fredga. 1984. Factors affecting reproductive output of the goldeneye duck *Bucephala clangula*. *Journal of Animal Ecology* 53: 679–692.

Dow, H., and S. Fredga. 1985. Selection of nest sites by a hole-nesting duck, the goldeneye *Bucephala clangula*. *Ibis* 127: 16–30.

DuWors, M. R., C. S. Houston, and P. W. Brown. 1984. Survival of the common goldeneye banded at Emma Lake, Saskatchewan. *Journal of Field Ornithology* 55: 382–383.

Eriksson, M. O. G. 1976. Food and feeding habits of downy goldeneye *Bucephala clangula* (L.) ducklings. *Ornis Scandinavica* 7: 159–169.

Eriksson, M. O. G. 1978. Lake selection by goldeneye ducklings in relation to the abundance of food. *Wildfowl* 29: 81–85.

Eriksson, M. O. G. 1983. The role of fish in the selection of lakes by non-piscivorous ducks: mallard, teal, and goldeneye. *Wildfowl* 34: 27–32.

Eriksson, M. O. G., and M. Andersson. 1982. Nest parasitism and hatching success in a population of goldeneyes. *Bird Study* 29: 49–54.

Foley, R. E., and G. R. Batcheller. 1988. Organochlorine contaminants in common goldeneye wintering on the Niagara River. *Journal of Wildlife Management* 52: 441–445.

Fredga, S., and H. Dow. 1984. Annual variation in the reproductive performance of goldeneyes. *Wildfowl* 34: 120–126.

Gibbs, R. M. 1961. Breeding ecology of the common goldeneye, *Bucephala clangula americana*, in Maine. MS thesis, University of Maine, Orono.

Gibbs, R. M. 1962. Juvenile mortality of the common goldeneye, *Bucephala clangula americana*, in Maine. *Maine Field Naturalist* 18: 67–68.

Grenquist, P. 1963. Hatching losses of common goldeneye. Pp. 685–689, *Proceedings XIII International Ornithological Congress, 1962*, vol. 2.

Johnson, L. L. 1967. The common goldeneye duck and the role of nesting boxes in its management in north-central Minnesota. *Journal of the Minnesota Academy of Science* 34: 110–113.

Jones, J. J., and R. D. Drobney. 1986. Winter feeding ecology of scaup and common goldeneye in Michigan. *Journal of Wildlife Management* 50: 446–452.

Lind, H. 1959. Studies on courtship and copulatory behavior in the goldeneye [*Bucephala clangula* (L.)]. *Dansk Ornithologisk Forenings Tidsskrift* 53: 177–219.

Mallory, M. L. 1991. Acid precipitation, female quality, and parental investment of common goldeneyes. MS thesis, Carleton University, Ottawa, ON.

Mallory, M. L., and P. J. Weatherhead. 1993a. Incubation rhythms and mass loss of common goldeneyes. *Condor* 95: 849–859.

Mallory, M. L., and P. J. Weatherhead. 1993b. Responses of nesting mergansers to parasitic common goldeneye eggs. *Animal Behaviour* 46: 1226–1228.

Mallory, M. L., D. K. McNicol, and P. J. Weatherhead. 1994. Habitat quality and reproductive effort of common goldeneyes nesting near Sudbury, Canada. *Journal of Wildlife Management* 58: 552–560.

Milonoff, M., H. Pöysä, and P. Runko. 1998. Factors affecting clutch size and duckling survival in the common goldeneye *Bucephala clangula. Wildlife Biology* 4: 73–80.

Milonoff, M., H. Pöysä, P. Runko, and V. Ruusila. 2004. Brood rearing costs affect future reproduction in the precocial common goldeneye *Bucephala clangula. Journal of Avian Biology* 35: 344–351.

Nilsson, L. 1969. The behaviour of the goldeneye *Bucephala clangula* in winter. *Vår Fågelvärld* 28: 199–210. (In Swedish, with English summary.)

Nilsson, L. 1971. Migration, nest-site tenacity, and longevity of Swedish goldeneye *Bucephala clangula. Vår Fågelvärld* 30: 180–183. (In Swedish, with English summary.)

Paasivaara, A., and H. Pöysä. 2004. Mortality of common goldeneye (*Bucephala clangula*) broods in relation to predation risk by northern pike (*Esox lucius*). *Annales Zoologici Fennici* 41: 513–523.

Paasivaara, A., and H. Pöysä. 2007. Survival of common goldeneye *Bucephala clangula* ducklings in relation to weather, timing of breeding, brood size, and female condition. *Journal of Avian Biology* 38: 144–152.

Paasivaara, A., J. Rutila, H. Pöysä, and P. Runko. 2010. Do parasitic common goldeneye *Bucephala clangula* females choose nests on the basis of host traits or nest site traits? *Journal of Avian Biology* 41: 662–671.

Pöysä, H. 2003. Low host recognition tendency revealed by experimentally induced parasitic egg laying in the common goldeneye (*Bucephala clangula*). *Canadian Journal of Zoology* 81: 1561–1565.

Pöysä, H., and S. Pöysä. 2002. Nest-site limitation and density dependence of reproductive output in the common goldeneye *Bucephala clangula:* Implications for the management of cavity-nesting birds. *Journal of Applied Ecology* 39: 502–510.

Pöysä, H., K. Lindblom, J. Rutila, and J. Sorjonen. 2009. Reliability of egg morphology to detect conspecific brood parasitism in goldeneyes *Bucephala clangula* examined using protein fingerprinting. *Journal of Avian Biology* 40: 453–456.

Pöysä, H., K. Lindblom, J. Rutila, and J. Sorjonen. 2010. Response of parasitically laying goldeneyes to experimental nest predation. *Animal Behaviour* 80: 881–86.

Pöysä, H., M. Milonoff, V. Ruusila, and J. Virtanen. 1999. Nest-site selection in relation to habitat edge: Experiments in the common goldeneye. *Journal of Avian Biology* 30: 79–84.

Prince, H. H. 1965. The breeding ecology of wood duck (*Aix sponsa* L.) and common goldeneye (*Bucephala clangula* L.) in central New Brunswick. MS thesis, University of New Brunswick, Fredericton.

Prince, H. H. 1968. Nest sites used by wood ducks and common goldeneyes in New Brunswick. *Journal of Wildlife Management* 32: 489–500.

Ruusila, V., H. Pöysä, and P. Runko. 2001. Costs and benefits of female-biased natal philopatry in the common goldeneye. *Behavioral Ecology* 12: 686–690.

Savard, J.-P. L, and M. Robert. 2007. Use of nest boxes by goldeneyes in eastern North America. *Wilson Journal of Ornithology* 119: 28–34.

Sayler, R. D., and A. D. Afton. 1981. Ecological aspects of common goldeneyes *Bucephala clangula* wintering on the upper Mississippi River. *Ornis Scandinavica* 12: 99–108.

Schmidt, J. H., E. J. Taylor, and E. A. Rexstad. 2005. Incubation behaviors and patterns of nest attendance in common goldeneyes in interior Alaska. *Condor* 107: 167–172.

Schmidt, J. H., E. J. Taylor, and E. A. Rexstad. 2006. Survival of common goldeneye ducklings in interior Alaska. *Journal of Wildlife Management* 70: 792–798.

Siren, M. 1951. Increasing the goldeneye population with nest boxes. *Suomen Riista* (*Papers on Game Research*) 6: 83–101, 189–190. (In Finnish, with English summary.)

Siren, M. 1952. Studies on the breeding biology of the goldeneye. *Suomen Riista* 8: 101–111. (In Finnish, with English summary.)

Siren, M. 1957a. On the faithfulness of goldeneye to its nesting region and nesting site. *Suomen Riista* 11: 130–133. (In Finnish, with English summary.)

Siren, M. 1957b. How goldeneye broods can be tied to one's own game grounds. *Suomen Riista* 11: 59–64. (In Finnish, with English summary.)

Wayland, M., and D. K. McNicol. 1994. Movements and survival of common goldeneye broods near Sudbury, Ontario, Canada. *Canadian Journal of Zoology* 72: 1252–1259.

Zicus, M. C. 1990a. Renesting by a common goldeneye. *Journal of Field Ornithology* 61: 245–248.

Zicus, M. C., and M. R. Riggs. 1996. Change in body mass of female common goldeneyes during nesting and brood rearing. *Wilson Bulletin* 108: 61–71.

Zicus, M. C., and S. K. Hennes. 1989. Nest prospecting by common goldeneyes. *Condor* 91: 807–812.

Zicus, M. C., and S. K. Hennes. 1993. Diurnal time budgets of breeding common goldeneyes. *Wilson Bulletin* 105: 680–685.

Zicus, M. C., and S. K. Hennes. 1994. Diurnal time budgets of common goldeneye brood hens. *Wilson Bulletin* 106: 549–554.

Zicus, M. C., S. K. Hennes, and M. R. Riggs. 1995. Common goldeneye nest attendance patterns. *Condor* 97: 461–472.

Multiple Merganser Species

Gregory, R. D., S. P. Carter, and S. R. Baillie. 1997. Abundance, distribution, and habitat use of breeding goosanders *Mergus merganser* and red-breasted mergansers *Mergus serrator* on British rivers. *Bird Study* 44: 1–12.

Miegs, R. C., and C. A. Rieck. 1967. Mergansers and trout in Washington. Pp. 306–318, *Proceedings 47th Annual Conference, Western Association of State Game and Fish Commissioners*.

Pearce, J. M., K. G. McCracken, T. K. Christensen, and Y. N. Zhuravlev. 2009. Migratory patterns and population structure among breeding and wintering red-breasted mergansers (*Mergus serrator*) and common mergansers (*M. merganser*). *Auk* 126: 784–798.

Sjöberg, K. 1988. Food selection, food-seeking patterns, and hunting success of captive goosanders *Mergus merganser* and red-breasted mergansers *Mergus serrator* in relation to the behavior of their prey. *Ibis* 130: 79–93.

Hooded Merganser

Barbour, D. B., and A. R. DeGrange. 1982. Communal roosting in wintering hooded mergansers. *Journal of Field Ornithology* 53: 279–280.

Bouvier, J. M. 1974. Breeding biology of the hooded merganser in southwestern Québec, including interactions with common goldeneyes and wood ducks. *Canadian Field-Naturalist* 88: 323–330.

Clover, P. C. 1981. Breeding of hooded merganser in Alfalfa County, Oklahoma. *Bulletin of the Oklahoma Ornithological Society* 14: 28–29.

Davis, S., and P. Capobianco. 2006. The hooded merganser: A preliminary look at growth in numbers in the United States as demonstrated in the Christmas Bird Count database. *American Birds* 60: 27–33.

Doty, H. A., F. B. Lee, A. D. Kruse, J. W. Matthews, J. R. Foster, and P. M. Arnold. 1984. Wood duck and hooded merganser nesting on Arrowwood NWR, North Dakota. *Journal of Wildlife Management* 48: 577–580.

Dugger, B. D., K. M. Dugger, and L. H. Fredrickson. 1999. Annual survival rates of female hooded mergansers and wood ducks in southeastern Missouri. *Wilson Bulletin* 111: 1–6.

Dugger, B. D., L. C. Bollmann, and L. H. Fredrickson. 1999. Response of female hooded mergansers to eggs of an interspecific parasite. *Auk* 116: 269–273.

Fendley, T. T. 1980. Incubating wood duck and hooded merganser hens killed by black rat snakes. *Wilson Bulletin* 92: 526–527.

Fournier, M. A., and J. E. Hines. 1996. Changed status of the hooded merganser, *Lophodytes cucullatus*, in the Yellowknife area, Northwest Territories. *Canadian Field-Naturalist* 110: 713–714.

Hansen, J. L. 1971. The role of nest boxes in the management of the wood duck on Mingo National Wildlife Refuge. MS thesis, University of Missouri, Columbia.

Heusmann, H. W., T. J. Early, and B. J. Nikula. 2000. Evidence of an increasing hooded merganser population in Massachusetts. *Wilson Bulletin* 112: 413–415.

Johnsgard, P. A. 1961. The sexual behavior and systematic position of the hooded merganser. *Wilson Bulletin* 73: 227–236. http://digitalcommons.unl.edu/biosciornithology/9

Kingery, H. E. 1999. Hooded merganser breeding in Colorado. *Journal of the Colorado Field Ornithologists* 33: 180–182.

Kitchen, D. W., and G. S. Hunt. 1969. Brood habitat of the hooded merganser. *Journal of Wildlife Management* 33: 605–609.

Mallory, M. L., A. Taverner, B. Bower, and D. Crook. 2002. Wood duck and hooded merganser breeding success in nest boxes in Ontario. *Wildlife Society Bulletin* 30: 310–316.

Mallory, M. L., H. G. Lumsden, and R. A. Walton. 1993. Nesting habits of hooded mergansers *Mergus cucullatus* in northeastern Ontario. *Wildfowl* 44: 101–107.

McGilvrey, F. B. 1966b. Nesting of hooded mergansers on the Patuxent Wildlife Research Center, Laurel, Maryland. *Auk* 83: 477–479.

Pandolfino, E. R., J. Kwolek, and K. Kreitinger. 2006. Expansion of the breeding range of the hooded merganser within California. *Western Birds* 37: 228–236.

Pearce, J. M., P. Blums, and M. S. Lindberg. 2008. Site fidelity is an inconsistent determinant of population structure in the hooded merganser (*Lophodytes cucullatus*): Evidence from genetic, mark-recapture, and comparative data. *Auk* 125: 711–722.

Sénéchal, H., G. Gauthier, and J.-P. L. Savard. 2008. Nesting ecology of common goldeneyes and hooded mergansers in a boreal river system. *Wilson Journal of Ornithology* 120: 732–742.

Stallcup, R. 2002. Hooded merganser (*Lophodytes cucullatus*): recent breeding range expansions in California and potential for breeding in Nevada. *Great Basin Birds* 5: 45–46.

Zicus, M. C. 1990. Nesting biology of hooded mergansers using nest boxes. *Journal of Wildlife Management* 54: 637–643.

Zicus, M. C. 1997. Female hooded merganser body mass during nesting. *Condor* 99: 220–224.

Zicus, M. C., M. A. Briggs, and R. M. Pace, III. 1988. DDE, PCB, and mercury residues in Minnesota common goldeneye and hooded merganser eggs, 1981. *Canadian Journal of Zoology* 66: 1871–1876.

Red-breasted Merganser

Atkinson, K. M., and D. P. Hewitt. 1978. A note on the food consumption of the red-breasted merganser. *Wildfowl* 29: 87–91.

Bowles, W. F., Jr. 1980. Winter ecology of red-breasted mergansers on the Laguna Madre of Texas. MS thesis, Corpus Christi State University, Corpus Christi, TX.

Bur, J. T., M. A. Stapanian, G. Bernhardt, and M. W. Turner. 2008. Fall diets of red-breasted merganser (*Mergus serrator*) and walleye (*Sander vitreus*) in Sandusky Bay and adjacent waters of western Lake Erie. *American Midland Naturalist* 159: 147–161.

Craik, S. R., and R. D. Titman. 2008. Movements and habitat use by red-breasted merganser broods in eastern New Brunswick. *Wilson Journal of Ornithology* 120: 743–754.

Craik, S. R., and R. D. Titman. 2009. Nesting ecology of red-breasted mergansers in a common tern colony in eastern New Brunswick. *Waterbirds* 32: 282–292.

Craik, S. R., J-P. L. Savard, M. J. Richardson, and R. D. Titman. 2011. Foraging ecology of flightless male red-breasted mergansers in the Gulf of St. Lawrence, Canada. *Waterbirds* 34: 280–288.

Curth, P. 1954. *Der Mittelsager: Soziologie und Brutbiologie.* Die Neue Brehm-Bücherei, Wittenberg, Lutherstadt, Germany.

Des Lauriers, J. R., and B. H. Brattstrom. 1965. Cooperative feeding behavior in red-breasted mergansers. *Auk* 82: 639.

Emlen, S. T., and H. W. Ambrose, III. 1970. Feeding interactions of snowy egrets and red-breasted mergansers. *Auk* 87: 164–165.

Grenquist, P. 1958. A nesting box for the red-breasted merganser. (In Finnish, with English summary). *Suomen Riista* 12: 94–99.

Heinz, G. H., S. D. Haseltine, W. L. Reichel, and G. L. Hensler. 1983. Relationships of environmental contaminants to reproductive success in red-breasted mergansers *Mergus serrator* from Lake Michigan. *Environmental Pollution, Series A* 32: 211–232.

Hending, P., B. King, and R. Prytherch. 1963. Communal diving in turbid waters by red-breasted mergansers. *Wildfowl Trust Annual Report* 14: 172–173.

Hofmann, T., J. W. Chardine, and H. Blokpoel. 1997. First breeding record of red-breasted merganser, *Mergus serrator*, on Axel Heiberg Island, Northwest Territories. *Canadian Field-Naturalist* 111: 308–309.

Kahlert, J. 1993. Breeding biology of the red-breasted merganser (*Mergus serrator*). MS thesis, University of Århus, Denmark.

Kahlert, J. 1994. Effects of human disturbance on broods of red-breasted mergansers *Mergus serrator*. *Wildfowl* 45: 222–231.

Kahlert, J., M. Coupe, and F. Cooke. 1998. Winter segregation and timing of pair formation in red-breasted merganser *Mergus serrator*. *Wildfowl* 49: 161–172.

Munro, J. A., and W. A. Clemens. 1939. The food and feeding habits of the red-breasted merganser in British Columbia. *Journal of Wildlife Management* 3: 46–53.

Nilsson, L. 1965. Observations of the spring behaviour of the red-breasted merganser. *Vår Fågelvärld* 24: 244–256.

Ringlegen, H. 1951. Aus dem leben des Mittelsagers (*Mergus serrator* L.). *Vogelwelt* 72: 43–50.

van der Kloot, W., and M. J. Morse. 1975. A stochastic analysis of the display behavior of the red-breasted merganser (*Mergus serrator*). *Behaviour* 54: 181–216.

White, H. C. 1939. The food of *Mergus serrator* on the Margaree River, N.S. *Bulletin of the Fisheries Research Board of Canada* 4b: 309–311.

Young, A. D., and R. D. Titman. 1986. Costs and benefits to red-breasted mergansers nesting in tern and gull colonies. *Canadian Journal of Zoology* 64: 2339–2343.

Young, A. D., and R. D. Titman. 1988. Intraspecific nest parasitism in red-breasted mergansers. *Canadian Journal of Zoology* 66: 2454–2458.

American Merganser and Goosander

Alcorn, J. R. 1953. Food of the common merganser in Churchill County, Nevada. *Condor* 55: 151–152.

Anderson, B. W., and R. L. Timken. 1971. Age and sex characteristics of common mergansers. *Journal of Wildlife Management* 35: 388–393.

Anderson, B. W., and R. L. Timken. 1972. Sex and age ratios and weights of common mergansers. *Journal of Wildlife Management* 36: 1127–1133.

Anderson, B. W., M. G. Reeder, and R. L. Timken. 1974. Notes on the feeding behavior of the common merganser (*Mergus merganser*). *Condor* 76: 472–476.

Brown, B. T. 1990. Nesting common mergansers in Chihuahua, Mexico. *Southwestern Naturalist* 35: 88–89.

Coldwell, C. 1939. The feeding habits of American mergansers. *Canadian Field-Naturalist* 53: 53–55.

Cunningham, R. 1981. Nesting behavior of the common merganser. *Loon* 63: 188–190.

Eriksson, K., and J. Niittylä. 1985. Breeding performance of the goosander *Mergus merganser* in the archipelago of the Gulf of Finland. *Ornis Fennica* 62: 153–157.

Erskine, A. J. 1971. Growth, and annual cycles in weights, plumages and reproductive organs of goosanders in eastern Canada. *Ibis* 113: 42–58.

Erskine, A. J. 1972. *Populations, Movements, and Seasonal Distribution of Mergansers in Northern Cape Breton Island.* Report Series 17. Canadian Wildlife Service, Ottawa.

Feltham, M. J. 1995. Consumption of Atlantic salmon smolts and parr by goosanders: estimates from doubly labelled water measurements of captive birds released on two Scottish rivers. *Journal of Fish Biology* 46: 273–281.

Foreman, L. D. 1976. Observations of common merganser broods in northwestern California. *California Fish and Game* 62: 207–212.

Fritsch, L. F., and I. O. Buss. 1958. Food of the American merganser in Unakwik Inlet, Alaska. *Condor* 60: 410–411.

Gilroy N. 1909. Notes on the nesting of the goosander. *British Birds* 2: 400–405.

Grenquist, P. 1953. On the nesting of the goosander in bird-boxes. *Suomen Riista* 8: 49–59.

Heard, W. R., and M. R. Curd. 1959. Stomach contents of American mergansers, *Mergus merganser* Linnaeus, caught in gill nets set in Lake Carl Blackwell, Oklahoma. *Oklahoma Academy of Science Proceedings* 39: 197–200.

Huntington, E. H., and A. A. Roberts. 1959. *Food Habits of the Merganser in New Mexico.* Bulletin 9. New Mexico Department of Game and Fish, Santa Fe.

Kiff, L. F. 1989. Historical breeding records of the common merganser in the southeastern United States. *Wilson Bulletin* 101: 141–143.

Latta, W. C., and R. F. Sharkey. 1966. Feeding behavior of the American merganser in captivity. *Journal of Wildlife Management* 30: 17–23.

Mallory, M. L., and H. G. Lumsden. 1994. Notes on egg laying and incubation in the common merganser. *Wilson Bulletin* 106: 757–759.

McCaw, J. H., III, P. J. Zwank, and R. L. Steiner. 1996. Abundance, distribution, and behavior of common mergansers wintering on a reservoir in southern New Mexico. *Journal of Field Ornithology* 67: 669–679.

Miller, S. W. 1973. The common merganser: Its wintering distribution and predation in a warm water reservoir. MS thesis, Oklahoma State University, Stillwater.

Miller, S. W., and J. S. Barclay. 1973. Predation in warm water reservoirs by wintering common mergansers. *Proceedings of the Southeastern Association of Fish and Wildlife Agencies* 27: 243–252.

Munro, I. A., and W. A. Clemens. 1936. Food of the American merganser (*Mergus merganser americanus*) in British Columbia. Paper No. 3. *Canadian Field-Naturalist* 50(3): 34–36.

Munro, I. A., and W. A. Clemens. 1937. The American merganser in British Columbia and its relation to the fish population. *Biological Board of Canada, Bulletin* 55: 1–50.

Nilsson, L. 1966. The behaviour of the goosander (*Mergus merganser*) in the winter. *Vår Fågelvärld* 25: 148–160. (In Swedish, with English summary.)

Pearce, J. M., D. Zwiefelhofer, and N. Maryanski. 2009. Mechanisms of population heterogeneity among molting common mergansers on Kodiak Island, Alaska: Implications for genetic assessments of migratory connectivity. *Condor* 111: 283–293.

Salyer, J. C., II, and K. F. Lagler. 1940. The food and habits of the American merganser during winter in Michigan, considered in relation to fish management. *Journal of Wildlife Management* 4: 186–219.

Scheuhammer, A. M., A. H. K. Wong, and D. Bond. 1998. Mercury and selenium accumulation in common loons (*Gavia immer*) and common mergansers (*Mergus merganser*) from eastern Canada. *Environmental Toxicology and Chemistry* 17: 197–201.

Sjöberg, K. 1989. Time-related predator/prey interactions between birds and fish in a northern Swedish river. *Oecologia* 80: 1–10.

Stiller, J. C. 2011. Effects of common mergansers on hatchery-reared brown trout and spring movements of adult males in southeastern New York, USA. MS thesis, College of Environmental Science and Forestry, State University of New York, Syracuse.

Stutz, S. S. 1965. Size of common merganser broods. *Murrelet* 46: 47–48.

Timken, R. L., and B. W. Anderson. 1969. Food habits of common mergansers in the north-central United States. *Journal of Wildlife Management* 33: 87–91.

White, H. C. 1957. *Food and Natural History of Mergansers on Salmon Waters in the Maritime Provinces of Canada.* Bulletin 116. Fisheries Research Board of Canada, Ottawa, ON.

Wood, C. C. 1986. Dispersion of common merganser (*Mergus merganser*) breeding pairs in relation to the availability of juvenile Pacific salmon in Vancouver Island streams. *Canadian Journal of Zoology* 64: 756–765.

Wood, C. C. 1987a. Predation of juvenile Pacific salmon by the common merganser (*Mergus merganser*) on eastern Vancouver Island. 1: Predation during the seaward migration. *Canadian Journal of Fisheries and Aquatic Sciences* 44: 941–949.

Wood, C. C. 1987b. Predation of juvenile Pacific salmon by the common merganser (*Mergus merganser*) on eastern Vancouver Island. 2: Predation of stream-resident juvenile salmon by merganser broods. *Canadian Journal of Fisheries and Aquatic Sciences* 44: 950–959.

Wood, C. C., and C. M. Hand. 1985. Food-searching behaviour of the common merganser (*Mergus merganser*). 1: Functional responses to prey and predator density. *Canadian Journal of Zoology* 63: 1260–1270.

Other Anatidae

Bartmann. W. 1998. New observations on the Brazilian merganser. *Wildfowl* 39: 714.

Johnsgard, P. A. 1966. The biology and relationships of the torrent duck. *Wildfowl Trust Annual Report* 17: 66–74.

www.ingramcontent.com/pod-product-compliance
Lightning Source LLC
Chambersburg PA
CBHW080608270326

41928CB00016B/2964